微信小程序开发解析

翟东平 / 编著

清華大学出版社
北京

内容简介

本书从零基础开始，系统地讲解了与微信小程序开发相关的知识点。全书按照微信小程序学习的技术路线设计章节结构，先介绍微信小程序框架，然后讲解微信小程序组件与API，最后讲解微信小程序支付、小程序商城和小程序客服。

本书在进行知识讲解时，力求简单、高效、系统，使读者真正弄懂微信小程序的开发原理、使用场景和程序架构方案，并能融会贯通。同时，本书针对微信小程序开发的相关重要技术接口，出了示例程序和执行结果，以方便读者快速应用并解决问题。

本书既可以作为初学者学习微信小程序开发的教材，也可以作为实际开发人员的工具书，遇到技术难题时随时查阅，以快速解决各类应用问题。

图书在版编目（CIP）数据

微信小程序开发解析/翟东平编著. 一北京：清华大学出版社，2023.10
ISBN 978-7-302-64649-5

I. ①微… II. ①翟… III. ①移动终端一应用程序一程序设计 IV. ①TN929.53

中国国家版本馆 CIP 数据核字（2023）第 182367 号

责任编辑：贾小红
封面设计：姜 龙
版式设计：文森时代
责任校对：马军令
责任印制：沈 露

出版发行：清华大学出版社
网 址：https://www.tup.com.cn，https://www.wqxuetang.com
地 址：北京清华大学学研大厦 A 座 邮 编：100084
社 总 机：010-83470000 邮 购：010-62786544
投稿与读者服务：010-62776969，c-service@tup.tsinghua.edu.cn
质量反馈：010-62772015，zhiliang@tup.tsinghua.edu.cn
印 装 者：三河市君旺印务有限公司
经 销：全国新华书店
开 本：185mm×260mm 印 张：18.25 字 数：511 千字
版 次：2023 年 11 月第 1 版 印 次：2023 年 11 月第 1 次印刷
定 价：79.80 元

产品编号：096263-01

前　言

微信小程序是微信生态环境下的一个技术解决方案，其主要的优势在于“依附但不依赖于微信”。微信小程序可以独立成为一个移动端的应用程序，亦可以与企业微信、微信公众号相关联。

本书从零基础开始，系统地讲解了与微信小程序开发相关的知识点。全书按照微信小程序学习的技术路线设计章节结构，先介绍微信小程序框架，然后讲解微信小程序组件与 API，最后讲解微信小程序支付、小程序商城和小程序客服，使读者真正弄懂微信小程序的开发原理、使用场景和程序架构方案，并能融会贯通。

本书在进行知识讲解时，力求简单、高效、系统。

- 简单：本书力争使用简洁、准确、明快的语言，一语中的地讲解枯燥、抽象的知识点，以降低读者的学习门槛。
- 高效：本书在讲解相关知识点时，直接给出“最小程序集合”，针对某一知识点单独建立项目、单独讲解，以带领读者聚焦知识点，降低学习成本。
- 系统：本书编排上结合官方文献资料，重新梳理、调整官方文档，最大限度地降低读者阅读文档的障碍，尽量避免读者由于不了解“上下文语意”造成的困扰。

读者既可以将本书作为系统学习微信领域知识的教材，也可以将本书作为工具手册，遇到问题时如同查字典一般检索相关知识点。

注意，小程序支付、小程序商城、小程序客服等实际项目中必要的功能需求点，属于微信小程序开发高级阶段的知识，建议读者在学习完微信小程序基础知识后再系统学习。

扫描图书封底的“文泉云盘”二维码，读者可下载书中案例的源代码、教学 PPT 课件，并观看对应的教学微课。读者学习过程中遇到疑难问题，也可以关注笔者的微信，进行交流沟通。

为了方便读者系统地了解整个微信技术生态，除本书之外，笔者还将编写《企业微信开发详解》《移动支付开发实战》等书，力争打造一套完善的微信技术解决方案教程。

本书完稿之际，笔者心潮澎湃，千言万语难以表达内心的激动与振奋。衷心地希望通过我们不懈的努力，能使本书尽善尽美。然而，书中难免存在疏漏或瑕疵，诚恳地希望读者批评指正，我们携手共同打造精品。

翟东平

2023 年 10 月

目　录

第1章　概　　论

1.1　认识微信小程序

微信小程序是继微信公众号、企业微信之后，腾讯推出的微信体系中另一个重要组成部分。

微信小程序是一种不需要下载、安装即可使用的应用，它实现了人们触手可及的梦想，用户扫一扫二维码或者搜索小程序名称即可打开应用，体现了“用完即走”的理念，即用户不用关心安装太多应用的问题，应用随处可用，但又无须安装和卸载。

简而言之，微信小程序具有以下特性。

- 小程序是一个应用。
- 无须下载。
- 触手可及。
- 用完即走。
- 无须卸载。

微信小程序可以关联到公众号，也可以关联到企业微信，还可以相对独立，即不与其他系统关联。与微信公众号、企业微信一样，微信小程序可以实现支付。

相对而言，微信小程序的开发门槛较低，但这并不说明微信小程序开发更简单一些。相反，微信小程序的调试、上线、审核都必须按照腾讯的要求进行，并不像公众号、企业微信那样有较高的自由度。

与公众号和企业微信开发相比，微信小程序开发的优势有以下几点。

（1）微信小程序提供默认的样式，没有特殊需求，一定程度上可以减少用户界面（UI）设计工作。

（2）微信小程序中有众多组件可以使用，一定程度上可以提高开发效率。

（3）微信小程序有自己的单位，一定程度上可以减少布局适配的工作。

和 App 开发相比，微信小程序开发的优点是成本更低、开发速度更快，具有 B/S 开发基础的开发人员会很容易上手（腾讯为开发者解决了平台级别的整合，开发者无须单独考虑跨操作系统问题）；其缺点是对硬件的控制速度不及 App 开发。在微信小程序和 App 之间做技术选型时，要重点考虑这些因素。

在学习微信小程序开发之前，读者需要掌握 JavaScript、CSS 以及相关后台编程语言等知识。总体来说，微信小程序开发学习门槛较低，但需要提醒读者的是，软件开发领域中没有任何一项技术是简单的，若想学好任何一项软件技术，都需要付出辛苦的努力，学习微信小程序开发也不例外。

1.2　微信小程序官方文档

微信小程序和微信公众号一样，都需要通过微信公众平台页面（https://mp.weixin.qq.com/）进行登录，如图 1-1 所示。

图 1-1

将鼠标指针悬停到“小程序”图标上，页面将显示“设计”“运营”“社区”“服务市场”“小程序开发文档”“小游戏开发文档”6 个选项，如图 1-2 所示。

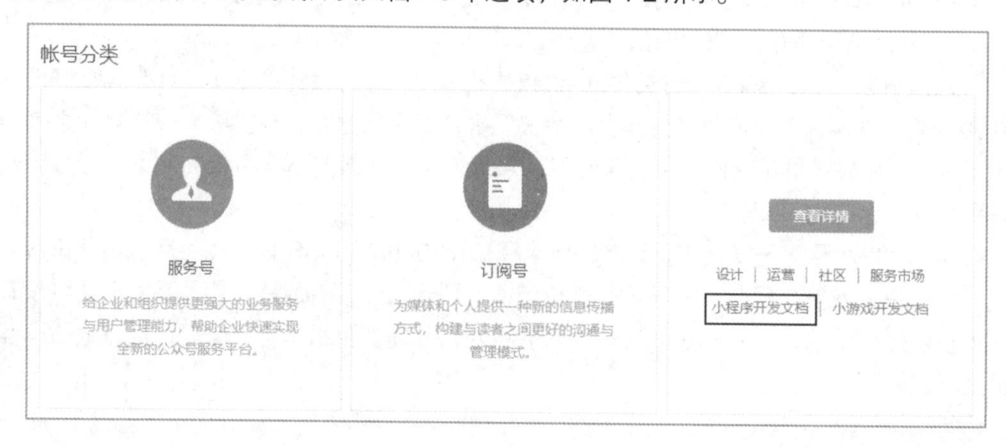

图 1-2

在图 1-2 中单击“小程序开发文档”链接，将打开微信小程序官方文档页面，如图 1-3 所示。默认显示的是“开发”页面，其中包含“指南”“框架”“组件”“API”“平台能力”“服务端”“工具”“云开发”“云托管”“更新日志”共 10 个选项卡，单击对应选项卡，可快速进入相应的文档页面。

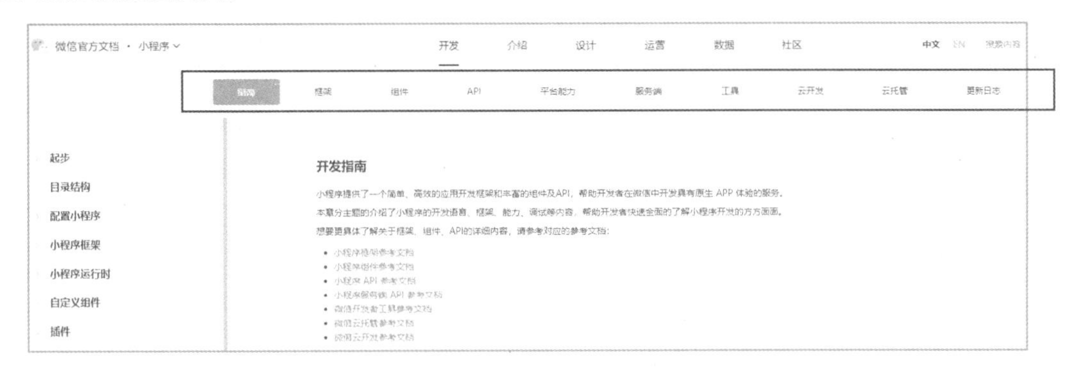

图 1-3

需提醒读者的是，微信小程序的版本更新速度较快，建议读者在学习时参考腾讯的最新官方文档。

1.3　下载微信小程序开发工具

在微信官方文档小程序开发页面中，选择上方的“工具”选项卡，打开的页面如图 1-4 所示。在左侧栏中单击“下载”链接，将打开小程序开发工具下载页面，如图 1-5 所示，其中可以看到稳定版、预发布版、开发版等多个版本。

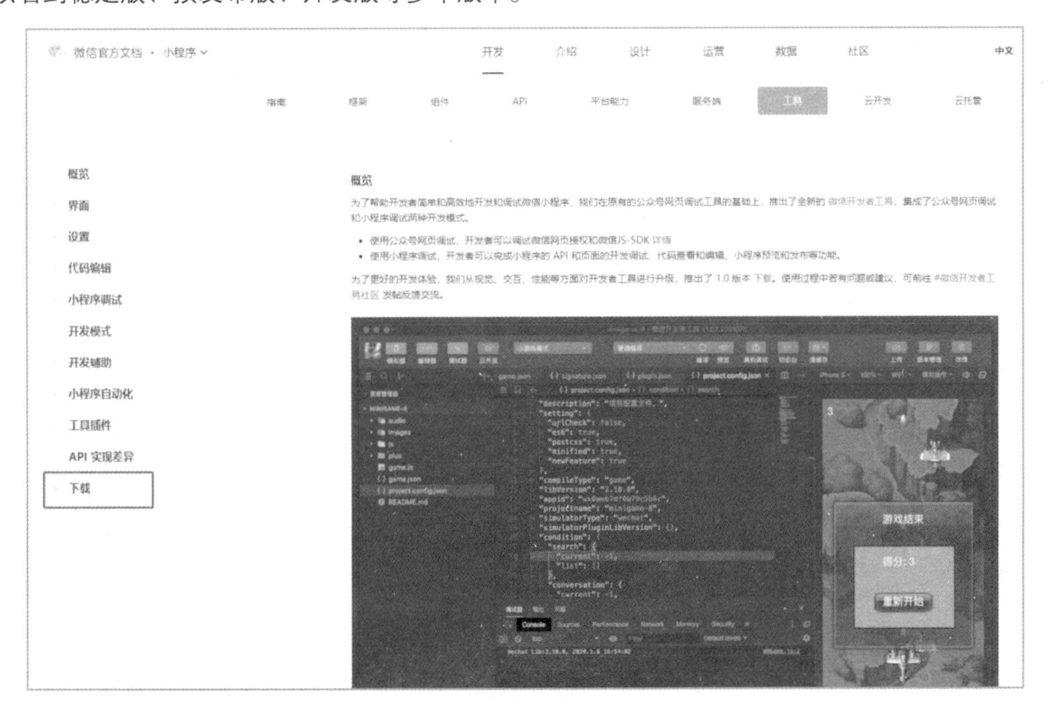

图 1-4

图 1-5

先下载适合的软件版本，然后根据提示步骤安装微信小程序开发工具即可。

1.4 使用微信小程序开发工具

打开安装好的微信小程序开发工具，如图 1-6 所示。

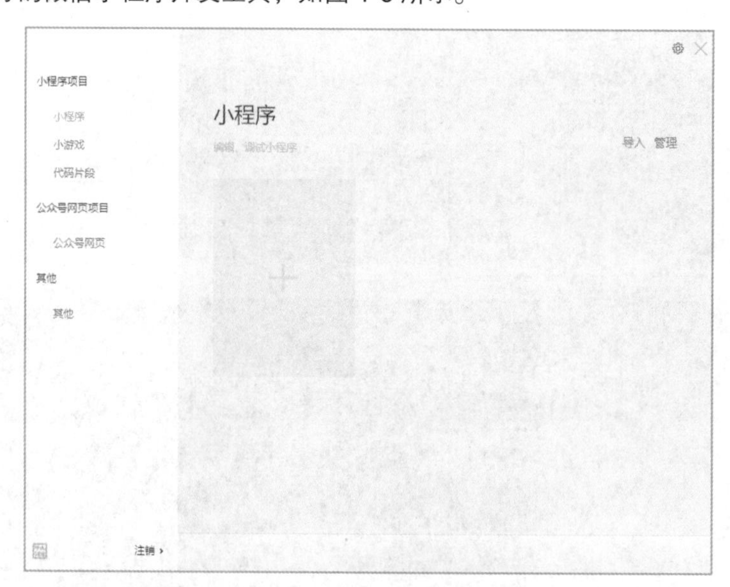

图 1-6

单击“+”按钮，新建一个小程序，并为其设置项目名称、目录、AppID、开发模式、后端服务、语言等信息，如图 1-7 所示。

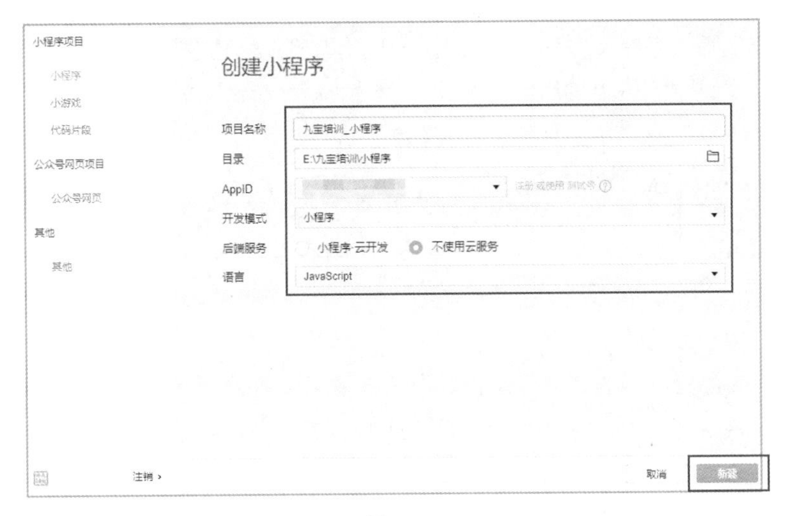

图 1-7

注意：

（1）AppID 指微信小程序 ID，读者可以根据需要注册新的账号或使用测试号。由于小程序支付需要使用正式账号，因此本书将使用微信小程序正式账号进行讲解。

（2）后端服务指微信小程序与后端连接的服务提供者，这里选中“不使用云服务”单选按钮。本书将引领读者自行搭建后台服务。

设置完毕后，单击“新建”按钮，将打开微信开发者工具，如图 1-8 所示，在此可对小程序项目进行开发和调试。页面上方是菜单栏，中间部分从左到右依次是模拟器、资源管理器和编辑器，编辑器的下方是调试器。

图 1-8

首先来认识一下菜单栏。左侧一组按钮用于显示或隐藏模拟器、编辑器、调试器，以及设置可视化和云开发，中间一组按钮用于开发过程中的程序编译、效果预览、真机调试和清理缓存，右侧一组按钮用于上传开发好的微信小程序并进行版本设置和管理，如图 1-9 所示。

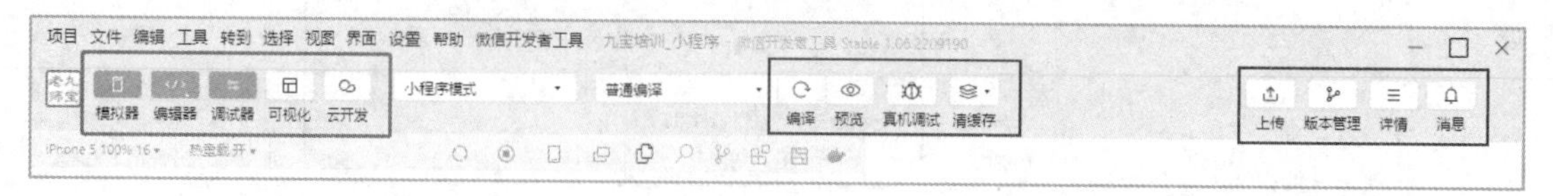

图 1-9

中间部分是资源管理器的树形目录，其中列出了相关的微信小程序文件。选中某个文件，在右侧编辑器中将会显示对应的小程序代码。调试程序可以使用 console.log，相关信息在右下方的调试器中显示，如图 1-10 所示。

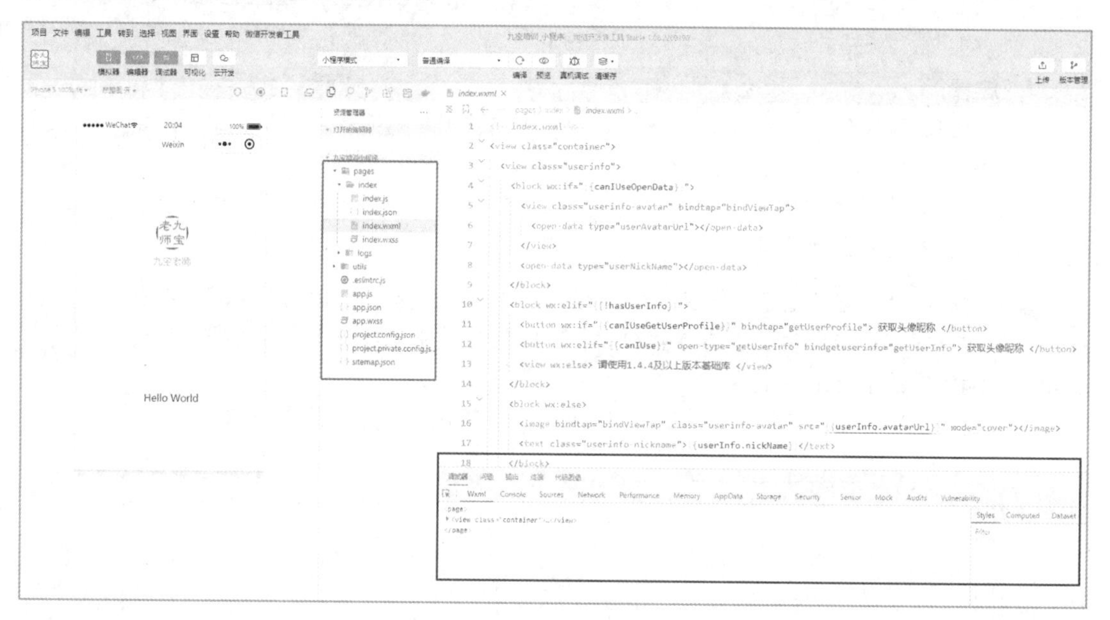

图 1-10

单击菜单栏中的“编译”按钮，可在左侧模拟器中预览小程序的运行效果。除此之外，还可以单击“预览”按钮，在弹出的面板中选择“二维码预览”或“自动预览”方式，使用真实的移动终端预览小程序效果。

其中，“二维码预览”方式需要读者使用手机微信客户端扫码观看预览效果，如图 1-11 所示。“自动预览”方式可将当前开发状态的微信小程序发布到开发人员的手机上，在手机端预览微信小程序的运行效果，如图 1-12 所示。

注意：

微信小程序开发工具需要关联微信账号。可以单击微信头像选择开发者，如图 1-13 所示。

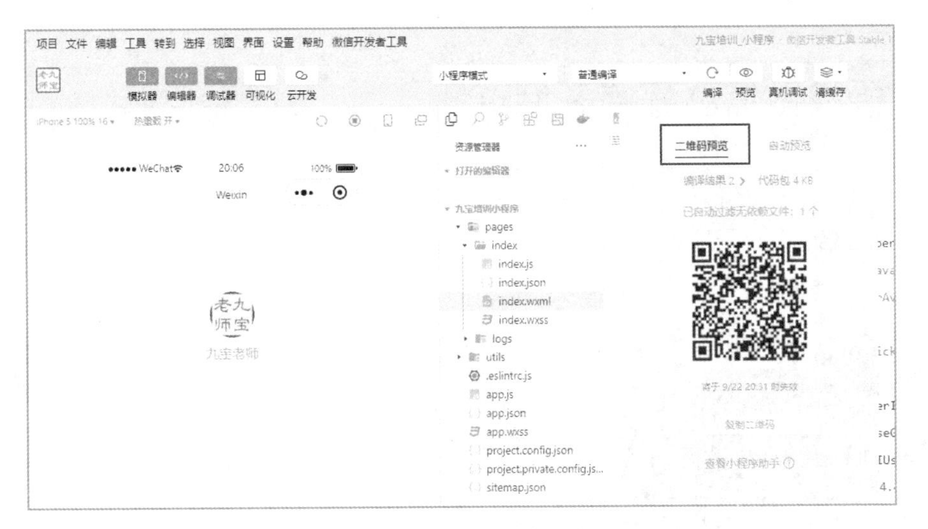

图 1-11

图 1-12

图 1-13

1.5　授权微信小程序开发者

在如图 1-14 所示的微信公众平台页面中，扫描右上方二维码，授权使用微信账号登录前面创建的“九宝培训”小程序。

图 1-14

登录系统后，在左侧栏中选择“成员管理”选项，在右侧页面的“项目成员”区中单击“编辑”右侧的下拉按钮，选择“添加成员”命令，如图 1-15 所示，添加新的用户。

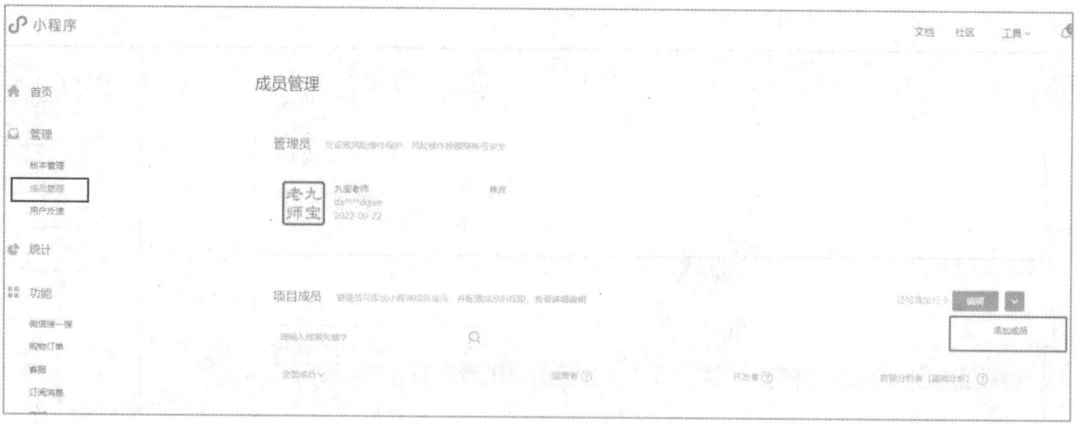

图 1-15

在“添加用户”页面中，先搜索要添加的微信号并为其设置权限，然后单击“确认添加”按钮，如图 1-16 所示。

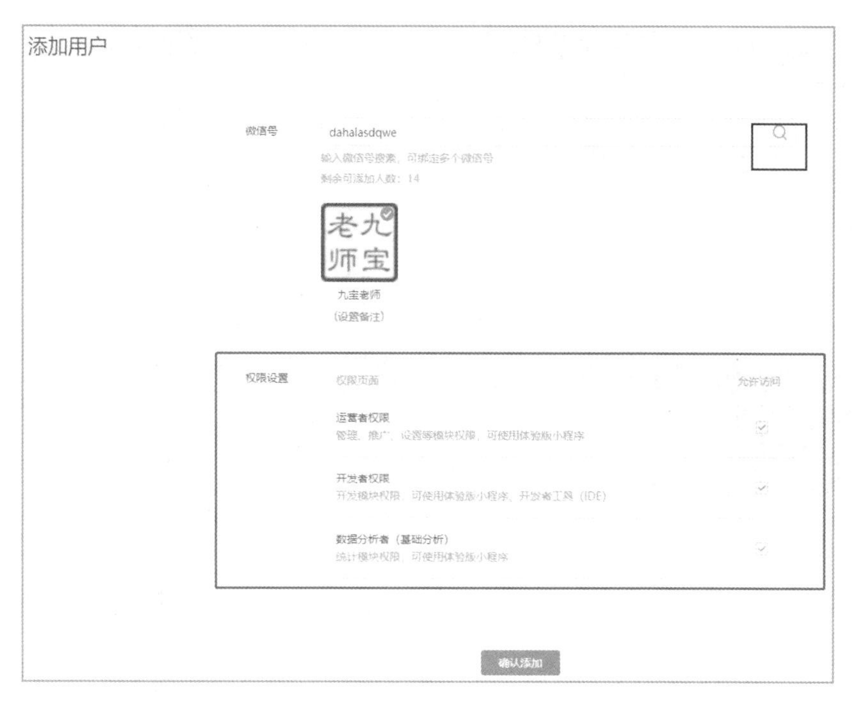

图 1-16

最多可以添加 15 个项目成员，添加成员后的页面如图 1-17 所示。

图 1-17

1.6　微信小程序开发工具特别关注点

1.6.1　真机调试功能

使用“预览”功能只能查看微信小程序的显示效果，即便是真机预览，也只能解决视图层面的显示问题。如果需要调试逻辑程序，就需要使用“真机调试”功能，如图 1-18 所示。

图 1-18

与“预览”功能相似，“真机调试”功能提供了“二维码真机调试”与“自动真机调试”两个选项，如图 1-19、图 1-20 所示。

采用自动真机调试时，手机端看到的效果如图 1-21 所示。此时，微信小程序开发工具将弹出一个对话框，如图 1-22 所示。

图 1-19

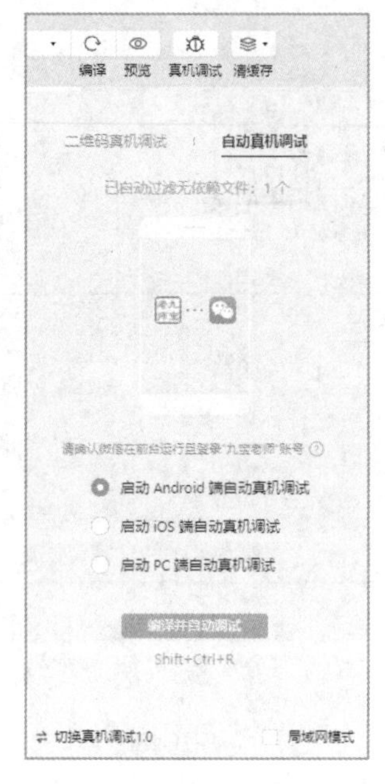

图 1-20

图 1-21

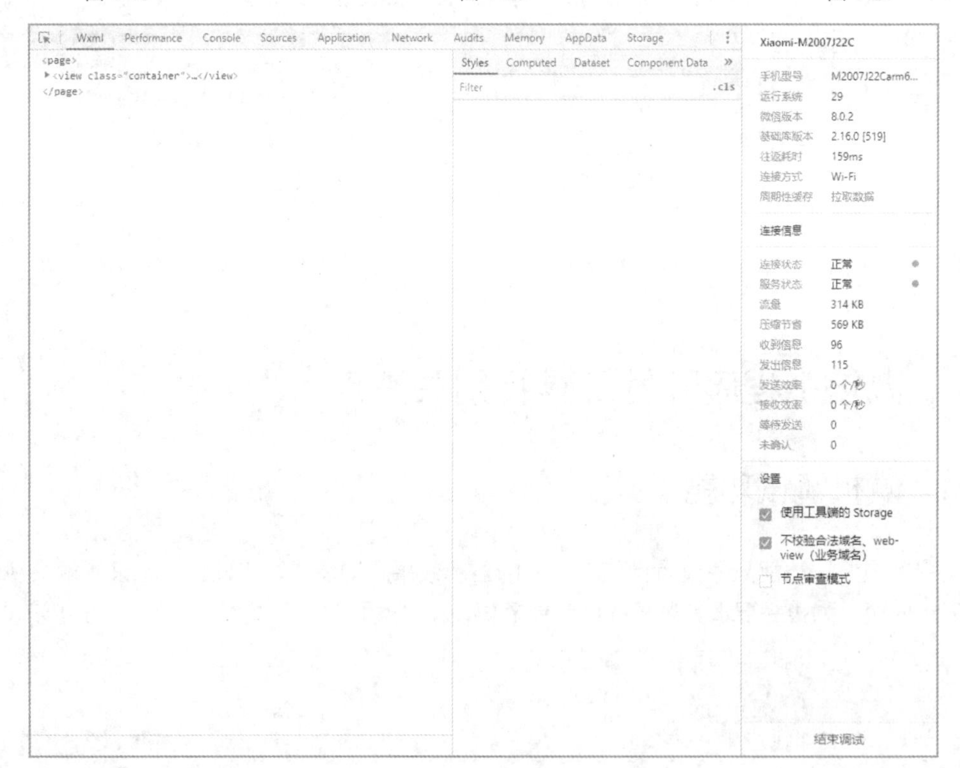

图 1-22

对于视图层调试，当鼠标指向某个 view 命令行时，手机端会提示对应的显示效果，如图 1-23、图 1-24 所示。

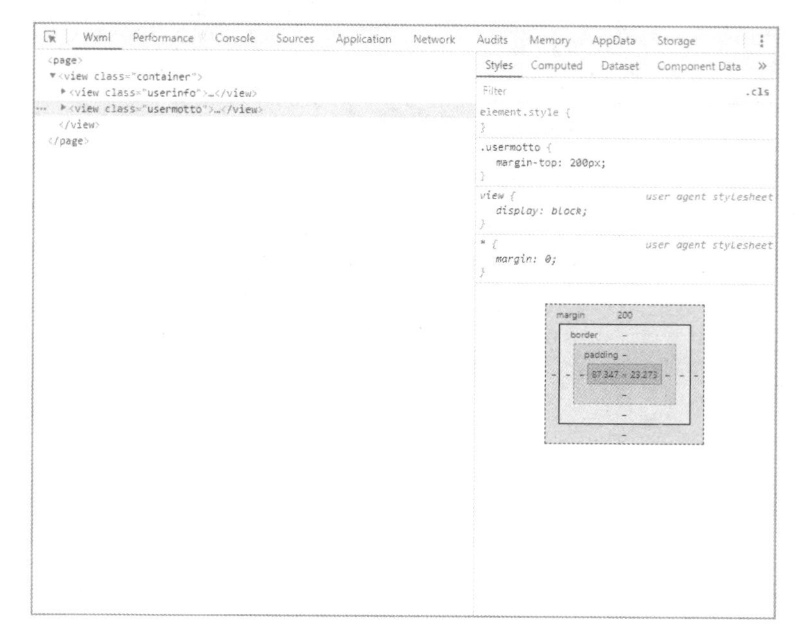

图 1-23

图 1-24

对于逻辑层调试，可以直接修改微信小程序中的变量，此时手机端将显示修改后的效果，如图 1-25、图 1-26 所示。

图 1-25

图 1-26

1.6.2　清除缓存功能

在微信小程序开发过程中，有时需要清理缓存。在菜单栏中单击“清缓存”按钮 ，如图 1-27 所示，在下拉菜单中选择一种清除缓存的方式，如图 1-28 所示。

图 1-27

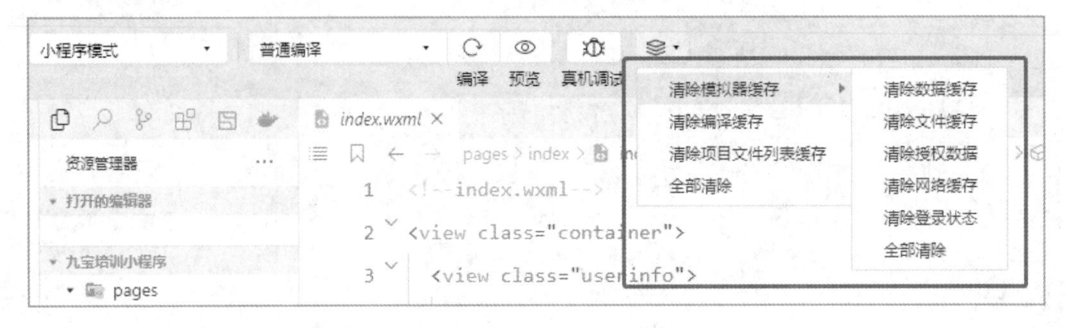

图 1-28

1.6.3 上传微信小程序

在正式发布微信小程序之前，需要先将其上传到腾讯官方提请审核。在菜单栏右侧单击“上传”按钮，如图 1-29 所示，然后确认上传小程序并设置版本信息，如图 1-30、图 1-31 所示。

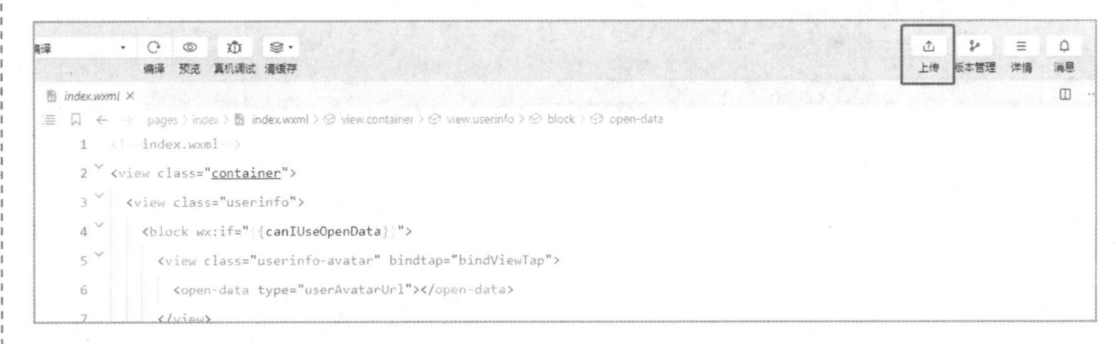

图 1-29

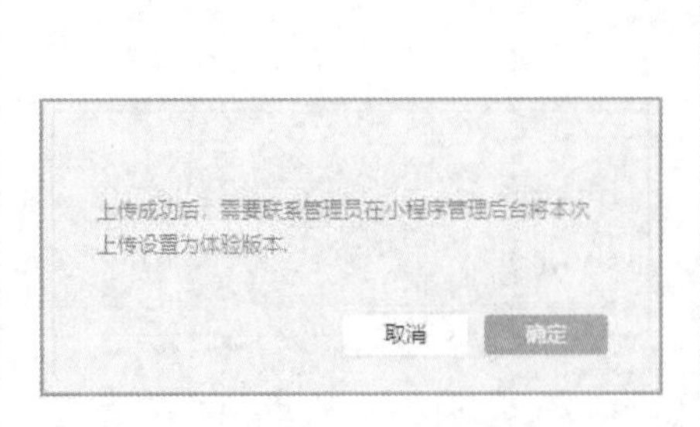

图 1-30

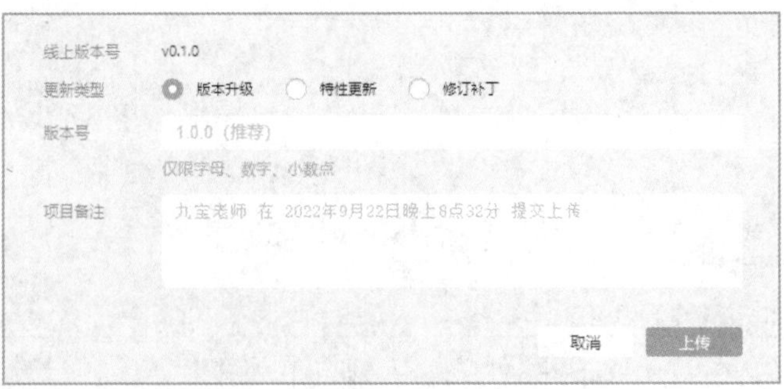

图 1-31

在微信小程序后台打开“版本管理”页面，可对已上传的小程序版本（如线上版本、审核版本、开发版本等）进行管理，如图 1-32 所示。例如，单击线上版本右下角的“详情”按钮，可回退该版本或暂停该版本服务，如图 1-33 所示。

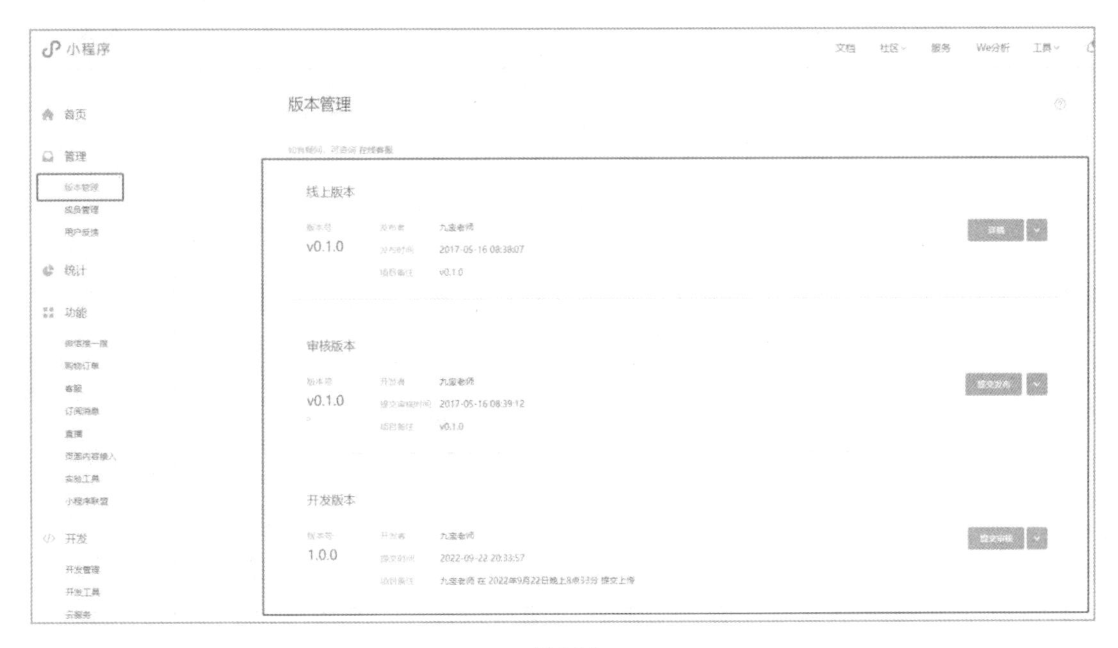

图 1-32

图 1-33

1.6.4　微信小程序体验版

要想使用微信小程序体验版，需在提交审核时先将其指定为体验版本，如图 1-34 所示。然后为其指定页面路径，单击“提交”按钮后体验版即可生效，如图 1-35、图 1-36 所示。

在微信小程序后台打开“版本管理”页面，在开发版本下方可看到“体验版”文字链接，如图 1-37 所示，单击该文字链接，将弹出体验版二维码，扫描二维码即可进入体验版，如图 1-38 所示。

图 1-34

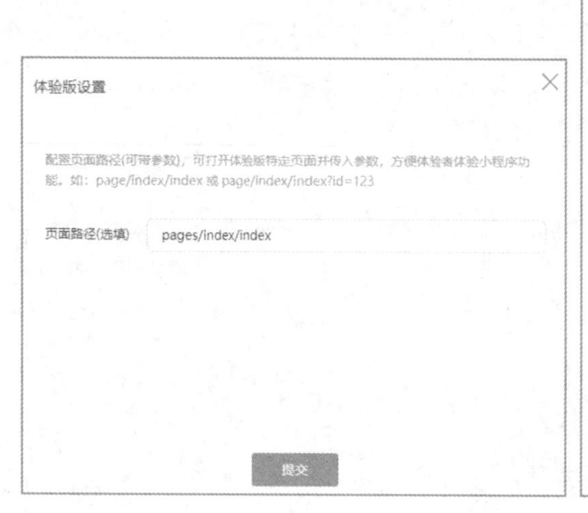

图 1-35

图 1-36

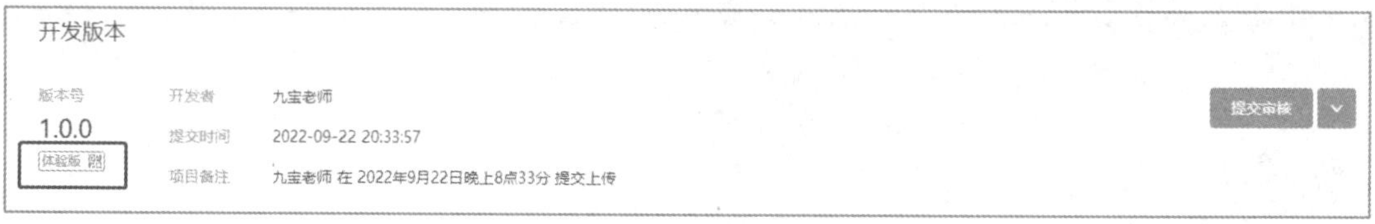

开发版本
版本号 1.0.0 体验版
开发者 九宝老师
提交时间 2022-09-22 20:33:57
项目备注 九宝老师 在 2022年9月22日晚上8点33分 提交上传
提交审核

图 1-37

图 1-38

需要注意的是，对于需要特殊权限才能体验的微信小程序，需在微信后台对体验成员进行特殊授权。

打开“成员管理”页面，在右侧“体验成员”区域单击“添加”按钮或“添加体验成员”文字链接，如图 1-39 所示，然后搜索并添加体验成员即可，如图 1-40 所示。

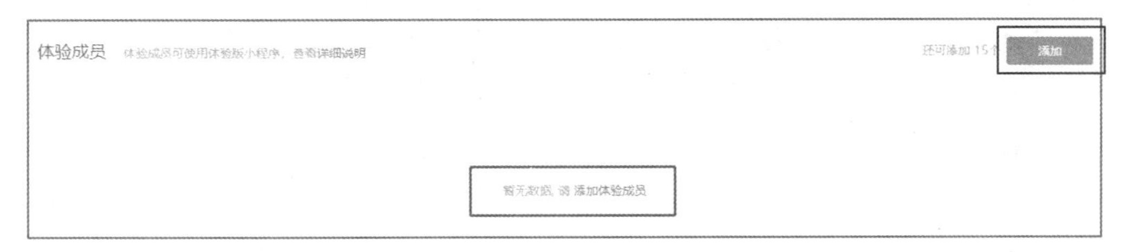

图 1-39

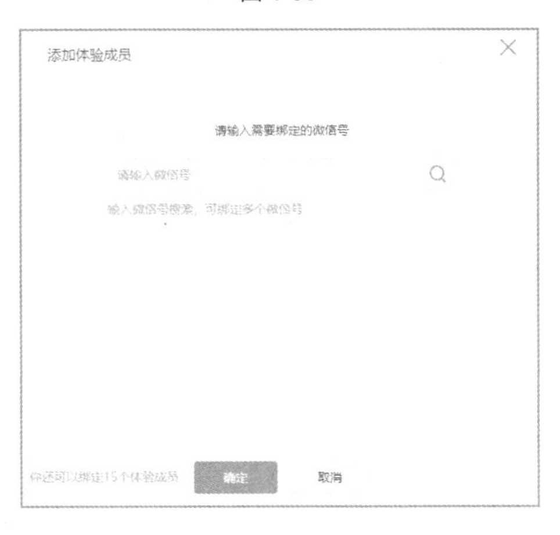

图 1-40

1.6.5　“详情”功能

在菜单栏中单击“详情”按钮，可快速获取微信小程序的相关配置信息，如图 1-41、图 1-42 所示。

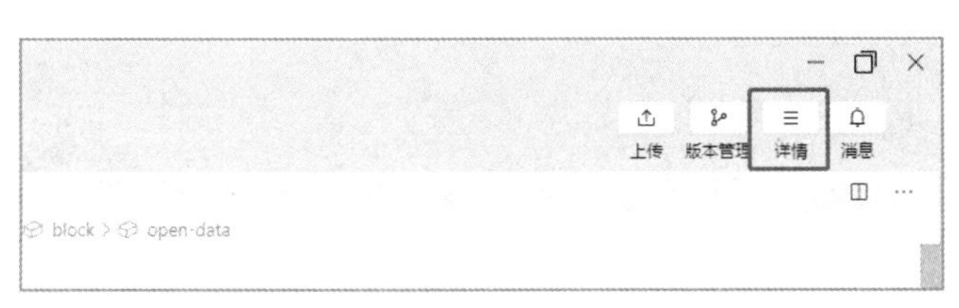

图 1-41

图 1-42

注意：

在“本地设置”选项卡中，通常需要选中“不校验合法域名、web-view（业务域名）、TLS

版本以及 HTTPS 证书”复选框，如图 1-43 所示；“项目配置”选项卡主要用来显示当前微信小程序账号的后台配置信息，重点是域名信息，如图 1-44 所示。

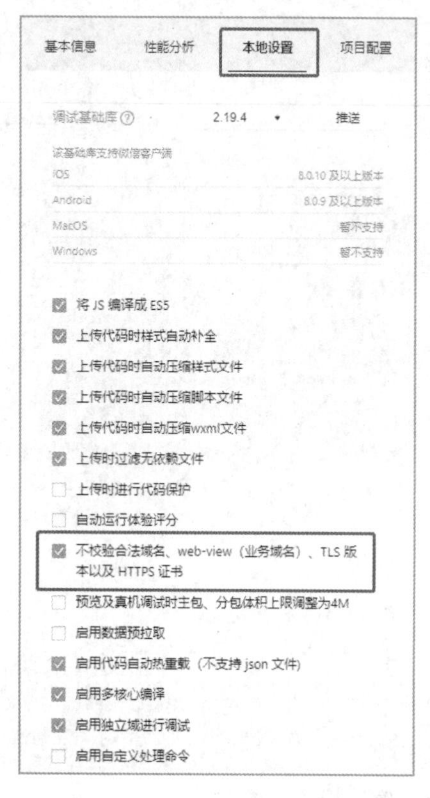

图 1-43

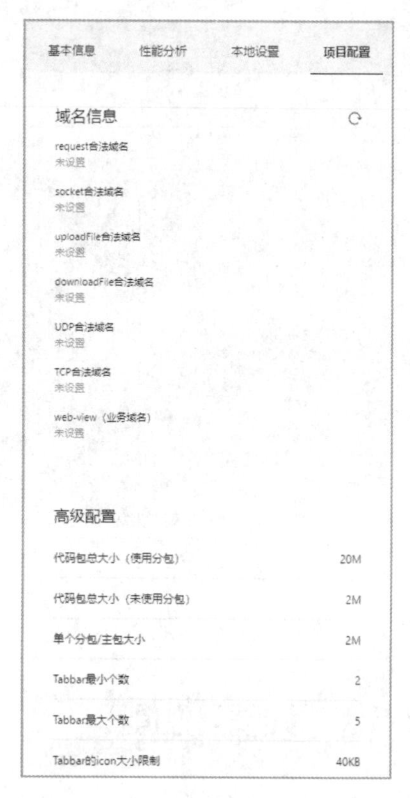

图 1-44

注意：

正式上线版本需要在小程序后台进行相应配置，包含配置 IP 白名单、服务器域名、数据预拉取等，如图 1-45～图 1-48 所示。

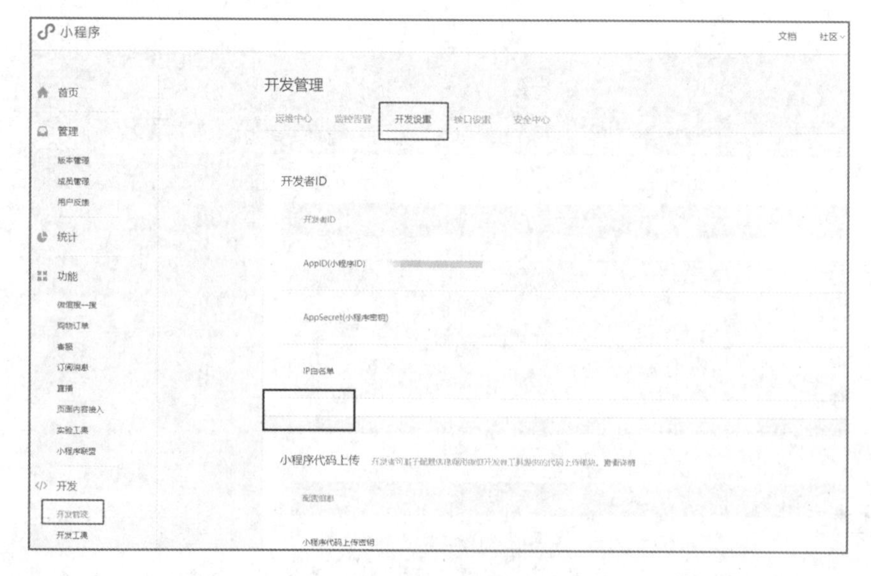

图 1-45

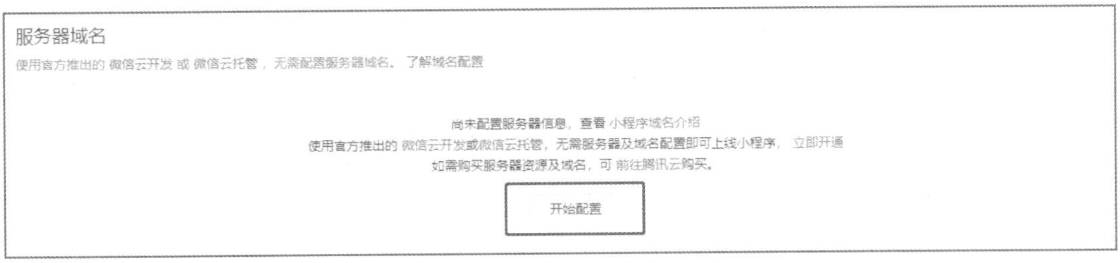

服务器域名

使用官方推出的 微信云开发 或 微信云托管，无需配置服务器域名。了解域名配置

尚未配置服务器信息，查看 小程序域名介绍

使用官方推出的 微信云开发或微信云托管，无需服务器及域名配置即可上线小程序，立即开通

如需购买服务器资源及域名，可 前往腾讯云购买。

开始配置

图 1-46

配置服务器域名

① 身份验证　② 配置服务器域名

可前往腾讯云购买服务器资源及域名。

request合法域名	以 https:// 开头。可填写200个域名，域名间请用；分割
socket合法域名	以 wss:// 开头。可填写200个域名，域名间请用；分割
uploadFile合法域名	以 https:// 开头。可填写200个域名，域名间请用；分割
downloadFile合法域名	以 https:// 开头。可填写200个域名，域名间请用；分割
udp合法域名	以 udp:// 开头。可填写200个域名，域名间请用；分割
tcp合法域名	以 tcp:// 开头。可填写200个域名，域名间请用；分割
DNS预解析域名	该域名无需填写协议头，最多填写5个域名，域名间请用；分割

保存并提交　取消

图 1-47

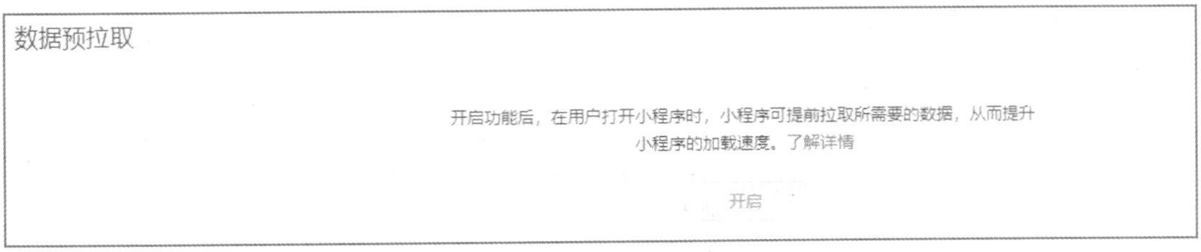

数据预拉取

开启功能后，在用户打开小程序时，小程序可提前拉取所需要的数据，从而提升小程序的加载速度。了解详情

开启

图 1-48

第 2 章　小程序基础知识

2.1　微信小程序开发与网页开发的区别

微信小程序开发与网页开发的区别主要有以下 6 点。

（1）微信小程序没有完整的浏览器对象，因此不能使用 jQuery、Zepto、Angular、requireJS 和 Vue。

（2）网页开发中，渲染线程和脚本线程是互斥的。微信小程序中，两者是分开的，分别运行在不同的线程中。

（3）微信小程序没有 DOM API 和 BOM API。

（4）微信小程序只需要针对 iOS 和 Android 微信客户端。

（5）不支持使用 eval 执行 JS 代码。

（6）不支持使用 new Function 创建函数。

微信小程序的运行环境如表 2-1 所示。

表 2-1

运 行 环 境	逻　辑　层	视　图　层
iOS	JavaScriptCore	WKWebView
Android	V8	chromium 定制内核
小程序开发者工具	NWJS	Chrome WebView

2.2　微信小程序的组成

微信小程序主要由以下 4 类文件组成。

（1）以.json 为后缀的 JSON 配置文件。

（2）以.wxml 为后缀的 WXML 模板文件。

（3）以.wxss 为后缀的 WXSS 样式文件。

（4）以.js 为后缀的 JS 脚本逻辑文件。

其中，JSON 配置文件用于提供微信小程序的配置信息，WXML 模板文件与 WXSS 样式文件用于实现微信小程序的视图层，JS 脚本逻辑文件用于实现微信小程序的逻辑层。这 4 个文件要求同名，且 WXML 模板文件与 JS 脚本逻辑文件是不可或缺的。

2.2.1　JSON 配置文件

微信默认创建的小程序根目录下包含 4 个 JSON 文件，分别是 app.json、project.config.

json、sitemap.json、index.json，如图 2-1 所示。下面依次介绍这些文件的作用。

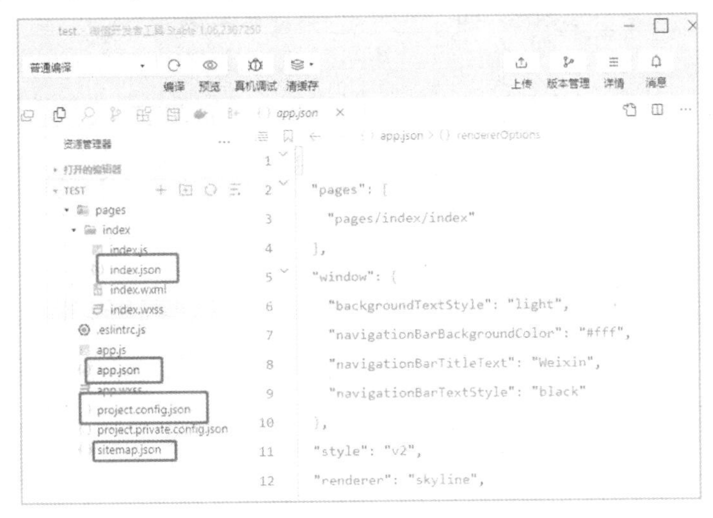

图 2-1

1．app.json 文件

app.json 是微信小程序的全局配置文件，主要用于配置小程序全局性的页面路径、界面表现、网络超时时间、底部 tab 栏等。默认创建的微信小程序的 app.json 代码如下。

```
{
  "pages":[
    "pages/index/index",
    "pages/logs/logs"
  ],
  "window":{
    "backgroundTextStyle":"light",
    "navigationBarBackgroundColor": "#fff",
    "navigationBarTitleText": "Weixin",
    "navigationBarTextStyle":"black"
  },
  "style": "v2",
  "sitemapLocation": "sitemap.json"
}
```

其中，各部分代码的含义及作用如下。

◆ pages 代码表明当前微信小程序由哪些页面组成，其中第一个页面代表当前微信小程序的首页。
◆ window 代码用于设置小程序的状态栏、导航条、标题、窗口背景色等。
◆ style 代码用于指定使用升级后的 WeUI 样式。
◆ sitemapLocation 用于指明 sitemap.json 的位置。

app.json 的相关配置项如表 2-2 所示。

表 2-2

属　性	类　型	必　填	描　述	最 低 版 本
entryPagePath	string	否	小程序默认启动首页	
pages	string[]	是	页面路径列表	

续表

属　　性	类　　型	必　　填	描　　述	最 低 版 本
window	Object	否	全局的默认窗口表现，具体参见表 2-3	
tabBar	Object	否	底部 tab 栏的表现，具体参见表 2-4	
networkTimeout	Object	否	网络超时时间，具体参见表 2-6	
debug	boolean	否	是否开启 debug 模式，默认关闭	
functionalPages	boolean	否	是否启用插件功能页，默认关闭	2.1.0
subpackages	Object[]	否	分包结构配置	1.7.3
workers	string	否	Worker 代码放置的目录	1.9.90
requiredBackgroundModes	string[]	否	需要在后台使用的能力，如「音乐播放」	
plugins	Object	否	使用到的插件	1.9.6
preloadRule	Object	否	分包预下载规则	2.3.0
resizable	boolean	否	PC 小程序是否支持用户任意改变窗口大小（包括最大化窗口）；iPad 小程序是否支持屏幕旋转。默认关闭	2.3.0
usingComponents	Object	否	全局自定义组件配置	开发者工具 1.02.1810190
permission	Object	否	小程序接口权限相关设置，具体参见表 2-7	微信客户端 7.0.0
sitemapLocation	string	是	指明 sitemap.json 的位置	
style	string	否	指定使用升级后的 WeUI 样式	2.8.0
useExtendedLib	Object	否	指定需要引用的扩展库	2.2.1
entranceDeclare	Object	否	微信消息用小程序打开	微信客户端 7.0.9
darkmode	boolean	否	小程序支持 DarkMode	2.11.0
themeLocation	string	否	指明 theme.json 的位置，darkmode 为 true 时为必填	开发者工具 1.03.2004271
lazyCodeLoading	string	否	配置自定义组件代码按需注入	2.11.1
singlePage	Object	否	单页模式相关配置，具体参见表 2-9	2.12.0

其中，window 的相关属性如表 2-3 所示。

表 2-3

属　　性	类　　型	默 认 值	描　　述	最 低 版 本
navigationBarBackgroundColor	HexColor	#000000	导航栏背景颜色，如 #000000	
navigationBarTextStyle	string	white	导航栏标题颜色，仅支持 black / white	
navigationBarTitleText	string		导航栏标题文字内容	
navigationStyle	string	default	导航栏样式，支持：default 为默认样式；custom 为自定义导航栏，只保留右上角胶囊按钮	iOS/Android 微信客户端 6.6.0，Windows 微信客户端不支持
backgroundColor	HexColor	#ffffff	窗口的背景色	
backgroundTextStyle	string	dark	下拉 loading 的样式，仅支持 dark/ light	
backgroundColorTop	string	#ffffff	顶部窗口的背景色，仅 iOS 支持	微信客户端 6.5.16
backgroundColorBottom	string	#ffffff	底部窗口的背景色，仅 iOS 支持	微信客户端 6.5.16
enablePullDownRefresh	boolean	false	是否开启全局的下拉刷新	
onReachBottomDistance	number	50	页面上拉触底事件触发时距页面底部的距离，单位为 px	
pageOrientation	string	portrait	屏幕旋转设置，支持 auto / portrait / landscape	2.4.0（auto）/ 2.5.0（landscape）

tabBar 以及 tabBar-list 的相关属性如表 2-4 和表 2-5 所示。

表 2-4

属　　性	类　　型	必　　填	默　认　值	描　　述	最 低 版 本
color	HexColor	是		tab 上的文字默认颜色，仅支持十六进制颜色	
selectedColor	HexColor	是		tab 上的文字选中时的颜色，仅支持十六进制颜色	
backgroundColor	HexColor	是		tab 的背景色，仅支持十六进制颜色	
borderStyle	string	否	black	tabBar 上边框的颜色，仅支持 black / white	
list	Array	是		tab 列表，包含最少 2 个、最多 5 个 tab	
position	string	否	bottom	tabBar 的位置，仅支持 bottom / top	
custom	boolean	否	false	自定义 tabBar	2.5.0

表 2-5

属　　性	类　　型	必　　填	说　　明
pagePath	string	是	页面路径，必须在 pages 中先定义
text	string	是	tab 上按钮文字
iconPath	string	否	图片路径，icon 大小限制为 40 KB，建议尺寸为 81px×81px，不支持网络图片。当 position 为 top 时，不显示 icon
selectedIconPath	string	否	选中时的图片路径，icon 大小限制为 40 KB，建议尺寸为 81px×81px，不支持网络图片。当 position 为 top 时，不显示 icon

networkTimeout 的相关属性如表 2-6 所示。

表 2-6

属　　性	类　　型	必　　填	默　认　值	说　　明
request	number	否	60000	wx.request 的超时时间，单位：毫秒
connectSocket	number	否	60000	wx.connectSocket 的超时时间，单位：毫秒
uploadFile	number	否	60000	wx.uploadFile 的超时时间，单位：毫秒
downloadFile	number	否	60000	wx.downloadFile 的超时时间，单位：毫秒

permission 以及 PermissionObject 的相关属性如表 2-7 和表 2-8 所示。

表 2-7

属　　性	类　　型	必　　填	默　认　值	描　　述
scope.userLocation	PermissionObject	否		位置相关权限声明

表 2-8

属　　性	类　　型	必　　填	默　认　值	说　　明
desc	string	是		小程序获取权限时展示的接口用途说明，最长 30 个字符

singlePage 的相关属性如表 2-9 所示。

表 2-9

属　　性	类　　型	必　　填	默　认　值	描　　述
navigationBarFit	String	否	默认自动调整，若原页面是自定义导航栏，则为 float，否则为 squeezed	导航栏与页面的相交状态，值为 float 时表示导航栏浮在页面上，与页面相交；值为 squeezed 时表示页面被导航栏挤压，与页面不相交

2．project.config.json 文件

project.config.json 是项目配置文件。

3．sitemap.json 文件

sitemap.json 文件用于配置小程序及其页面是否可被微信索引，文件内容为一个 JSON 对象。如果没有 sitemap.json，则默认为所有页面都允许被索引。

微信现已开放小程序内搜索，开发者可以通过 sitemap.json 进行配置，或者通过管理后台页面的收录开关来配置其小程序页面是否允许微信索引。当开发者允许微信索引时，微信会通过爬虫的形式，为小程序的页面内容建立索引。当用户的搜索词条触发该索引时，小程序的页面将可能展示在搜索结果中。

爬虫访问小程序内页面时，会携带特定的 user-agent：mpcrawler 及场景值：1129。需要注意的是，若小程序爬虫发现的页面数据和真实用户的数据呈现不一致，那么该页面将不会进入索引。

4．index.json 文件

index.json 文件可对 index 页面的窗口表现进行配置。页面中的配置项在当前页面会覆盖 app.json 的 window 中相同的配置项。文件内容为一个 JSON 对象。

Page 的相关配置项以及单页模式的相关属性如表 2-10、表 2-11 所示。

表 2-10

属　性	类　型	默 认 值	描　述	最 低 版 本
navigationBarBackgroundColor	HexColor	#000000	导航栏背景颜色，如 #000000	
navigationBarTextStyle	string	white	导航栏标题颜色，仅支持 black / white	
navigationBarTitleText	string		导航栏标题文字内容	
navigationStyle	string	default	导航栏样式，仅支持：default 为默认样式；custom 为自定义导航栏，只保留右上角胶囊按钮	iOS/Android 微信客户端 7.0.0，Windows 微信客户端不支持
backgroundColor	HexColor	#ffffff	窗口的背景色	
backgroundTextStyle	string	dark	下拉 loading 的样式，仅支持 dark/light	
backgroundColorTop	string	#ffffff	顶部窗口的背景色，仅 iOS 支持	微信客户端 6.5.16
backgroundColorBottom	string	#ffffff	底部窗口的背景色，仅 iOS 支持	微信客户端 6.5.16
enablePullDownRefresh	boolean	false	是否开启当前页面下拉刷新	
onReachBottomDistance	number	50	页面上拉触底事件触发时距页面底部距离，单位为 px	
pageOrientation	string	portrait	屏幕旋转设置，支持 auto/portrait/landscape	2.4.0（auto）/2.5.0（landscape）
disableScroll	boolean	false	设置为 true 则页面整体不能上下滚动。只在页面配置中有效，无法在 app.json 中设置	
usingComponents	Object	否	页面自定义组件配置	1.6.3
initialRenderingCache	string		页面初始渲染缓存配置	2.11.1
style	string	default	启用新版的组件样式	2.10.2
singlePage	Object	否	单页模式相关配置，具体参见表 2-11	2.12.0

表 2-11

属　　性	类　　型	必　　填	默 认 值	描　　述
navigationBarFit	String	否	默认自动调整，若原页面是自定义导航栏，则为 float，否则为 squeezed	导航栏与页面的相交状态，值为 float 时表示导航栏浮在页面上，与页面相交；值为 squeezed 时表示页面被导航栏挤压，与页面不相交

2.2.2　视图层

微信小程序的运行环境分为视图层与逻辑层，WXML 模板文件与 WXSS 样式文件工作在视图层，JS 脚本文件工作在逻辑层。

微信小程序的视图层和逻辑层分别由两个线程管理。视图层的界面使用 WebView 进行渲染；逻辑层采用 JSCore 线程运行 JS 脚本。一个微信小程序存在多个界面，所以视图层存在多个 WebView 线程。这两个线程的通信会经由微信客户端（下文中采用 Native 来代指微信客户端）中转，逻辑层发送网络请求也经由 Native 转发，如图 2-2 所示。

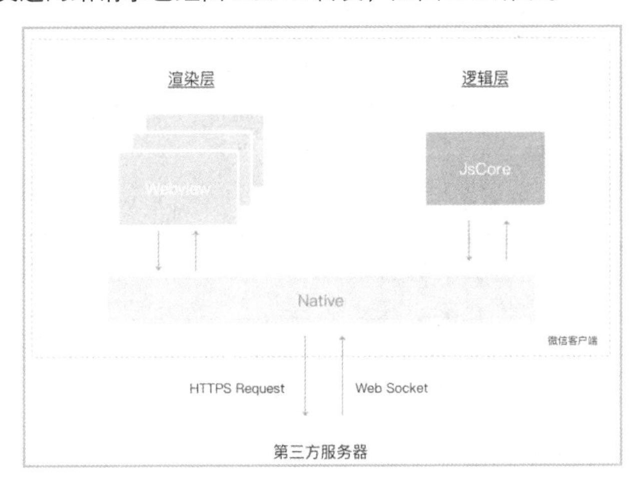

图 2-2

1. WXML 模板文件

微信小程序中的 WXML 模板文件相当于 Web 开发中的 HTML 文件。

下面先来看一个微信小程序实例，其中 WXML 模板文件的部分代码如下。

```
<!--index.wxml-->
<view class="container">
  <view class="userinfo">
    <block wx:if="{{canIUseOpenData}}" calss="userinfo-opendata">
      <view class="userinfo-avatar" bindtap="bindViewTap">
        <open-data type="userAvatarUrl"></open-data>
      </view>
      <open-data type="userNickName"></open-data>
    </block>
    <block wx:elif="{{!hasUserInfo}}">
      <button wx:if="{{canIUseGetUserProfile}}" bindtap="getUserProfile"> 获取头像昵称
</button>
      <button wx:elif="{{canIUse}}" open-type="getUserInfo"
bindgetuserinfo="getUserInfo"> 获取头像昵称 </button>
```

```
        <view wx:else> 请使用 1.4.4 及以上版本基础库 </view>
      </block>
      <block wx:else>
        <image bindtap="bindViewTap" class="userinfo-avatar"
src="{{userInfo.avatarUrl}}" mode="cover"></image>
        <text class="userinfo-nickname">{{userInfo.nickName}}</text>
      </block>
    </view>
    <view class="usermotto">
      <text class="user-motto">{{motto}}</text>
    </view>
  </view>
```

为了方便快速理解，读者可以将 view 理解成 Web 开发中的 div；实例程序中的 wx:if、wx:elif 和 wx:else 可以借鉴 Vue 的相关用法；bindtap 可以看作 onclick；class 与 Web 开发中的 class 含义一样。

2．WXSS 样式文件

WXSS 样式文件可以看作 Web 开发中的 CSS 文件。WXSS 具有 CSS 大部分的特性，同时做了一些扩充和修改，如新增了尺寸单位。WXSS 在底层支持新的尺寸单位——rpx，开发者可以免去换算的烦恼，把换算交给小程序底层来进行即可。换算采用的是浮点数运算，运算结果可能会和预期结果有一定偏差。

与 app.json、page.json 的概念相同，读者可以写一个 app.wxss 作为全局样式，作用于当前小程序的所有页面，而局部页面样式 page.wxss 仅对当前页面生效。WXSS 仅支持部分 CSS 选择器。

2.2.3 逻辑层

微信小程序的逻辑层主要用于实现相关业务的逻辑。

下面来看一个微信小程序实例，其中 JS 文件代码如下。

```
  // index.js
  // 获取应用实例
  const app = getApp()

  Page({
    data: {
      motto: 'Hello World',
      userInfo: {},
      hasUserInfo: false,
      canIUse: wx.canIUse('button.open-type.getUserInfo'),
      canIUseGetUserProfile: false,
      canIUseOpenData: wx.canIUse('open-data.type.userAvatarUrl') &&
wx.canIUse('open-data.type.userNickName') // 如需尝试获取用户信息可改为 false
    },
    // 事件处理函数
    bindViewTap() {
      wx.navigateTo({
        url: '../logs/logs'
      })
```

```
  },
  onLoad() {
    if (wx.getUserProfile) {
      this.setData({
        canIUseGetUserProfile: true
      })
    }
  },
  getUserProfile(e) {
    /* 推荐使用wx.getUserProfile获取用户信息,开发者每次通过该接口获取用户个人信息均需用户确认,
开发者应妥善保管用户快速填写的头像昵称，避免重复弹窗*/
    wx.getUserProfile({
      desc: '展示用户信息', // 声明获取用户个人信息后的用途，后续会展示在弹窗中，请谨慎填写
      success: (res) => {
        console.log(res)
        this.setData({
          userInfo: res.userInfo,
          hasUserInfo: true
        })
      }
    })
  },
  getUserInfo(e) {
    /* 不推荐使用getUserInfo获取用户信息，自2021年4月13日起，getUserInfo就不再弹窗，并直
接返回匿名的用户个人信息*/
    console.log(e)
    this.setData({
      userInfo: e.detail.userInfo,
      hasUserInfo: true
    })
  }
})
```

2.3 程序与页面

微信小程序分为 App 级别与 Page 级别。当微信小程序被下载到微信客户端后，按照先 App 后 Page 的顺序执行。

App 级别微信小程序的回调方法如下。

```
App({
  /**
   * 当小程序初始化完成时，会触发 onLaunch（全局只触发一次）
   */
  onLaunch: function () {

  },

  /**
   * 当小程序启动，或从后台进入前台显示时，会触发 onShow
   */
  onShow: function (options) {
```

```
  },

  /**
   * 当小程序从前台进入后台时，会触发 onHide
   */
  onHide: function () {

  },

  /**
   * 当小程序发生脚本错误，或者 API 调用失败时，会触发 onError 并带上错误信息
   */
  onError: function (msg) {

  }
})
```

Page 级别主要包含 JSON 配置文件、WXML 模板文件、WXSS 样式文件、JS 脚本逻辑文件。微信客户端先根据 JSON 配置生成一个界面，按照 JSON 的配置定义顶部的颜色和文字。之后，微信客户端装载 WXML 模板文件和 WXSS 样式文件。最后，微信客户端装载 JS 脚本逻辑文件。

Page 级别小程序的回调方法如下。

```
Page({
  /**
   * 页面的初始数据
   */
  data: {

  },

  /**
   * 生命周期函数--监听页面加载
   */
  onLoad: function (options) {

  },

  /**
   * 生命周期函数--监听页面初次渲染完成
   */
  onReady: function () {

  },

  /**
   * 生命周期函数--监听页面显示
   */
  onShow: function () {

  },

  /**
   * 生命周期函数--监听页面隐藏
   */
```

```
  onHide: function () {

  },

  /**
   * 生命周期函数--监听页面卸载
   */
  onUnload: function () {

  },

  /**
   * 页面相关事件处理函数--监听用户下拉动作
   */
  onPullDownRefresh: function () {

  },

  /**
   * 页面上拉触底事件的处理函数
   */
  onReachBottom: function () {

  },

  /**
   * 用户单击右上角分享
   */
  onShareAppMessage: function () {

  }
})
```

第 3 章　小程序框架

本章将系统讲解微信小程序框架的相关知识。

3.1　新建项目

新建一个微信小程序项目，删除多余的示例程序，保留最小程序集合。调整后的程序如图 3-1 所示。

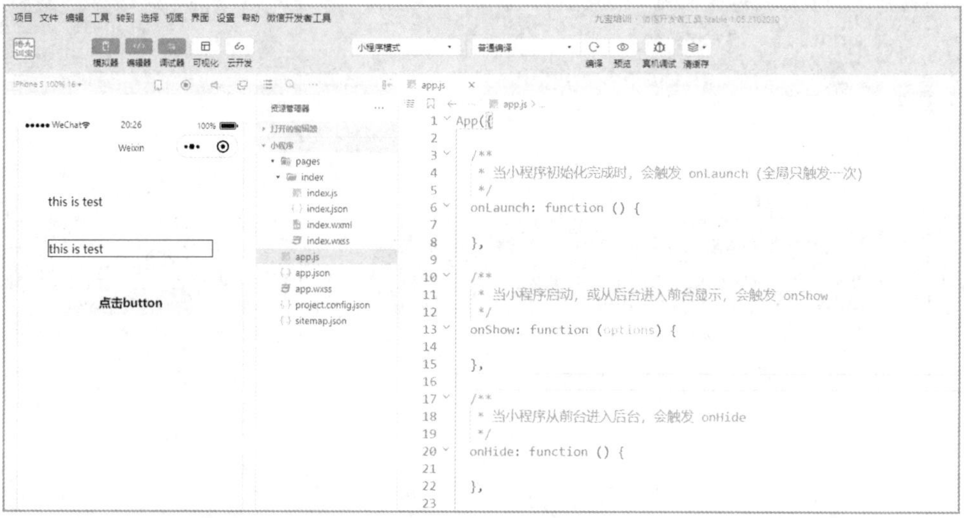

图 3-1

3.2　程序清单

以下程序清单将作为本书讲解微信小程序时的基础程序。如无特殊说明，这部分程序将作为后续章节所有新建程序的基础，并简称为“最小程序状态”，后文使用时不再赘述。

3.2.1　app.js

app.js 的示例代码如下。

```
App({
  /**
   * 当小程序初始化完成时，会触发 onLaunch（全局只触发一次）
   */
  onLaunch: function () {
```

```
},

  /**
   * 当小程序启动，或从后台进入前台显示时，会触发 onShow
   */
  onShow: function (options) {

  },

  /**
   * 当小程序从前台进入后台时，会触发 onHide
   */
  onHide: function () {

  },

  /**
   * 当小程序发生脚本错误，或者 API 调用失败时，会触发 onError 并带上错误信息
   */
  onError: function (msg) {

  }
})
```

3.2.2 app.json

app.json 的示例代码如下。

```
{
  "pages":[
    "pages/index/index"
  ],
  "window":{
    "backgroundTextStyle":"light",
    "navigationBarBackgroundColor": "#fff",
    "navigationBarTitleText": "Weixin",
    "navigationBarTextStyle":"black"
  },
  "style": "v2",
  "sitemapLocation": "sitemap.json"
}
```

3.2.3 index.js

index.js 的示例代码如下。

```
Page({
  /**
   * 页面的初始数据
   */
  data: {

  },

  /**
   * 生命周期函数--监听页面加载
   */
  onLoad: function (options) {
```

```
  },

  /**
   * 生命周期函数--监听页面初次渲染完成
   */
  onReady: function () {

  },

  /**
   * 生命周期函数--监听页面显示
   */
  onShow: function () {

  },

  /**
   * 生命周期函数--监听页面隐藏
   */
  onHide: function () {

  },

  /**
   * 生命周期函数--监听页面卸载
   */
  onUnload: function () {

  },

  /**
   * 页面相关事件处理函数--监听用户下拉动作
   */
  onPullDownRefresh: function () {

  },

  /**
   * 页面上拉触底事件的处理函数
   */
  onReachBottom: function () {

  },

  /**
   * 用户单击右上角分享
   */
  onShareAppMessage: function () {

  }
})
```

3.3 响应式数据绑定

构建项目的“最小程序状态”，修改 index.js 文件，代码如下。

```
Page({
  data: {
    arg:'this is test'
```

```
    },
    but_click(){
      this.setData({arg:'点击 but 修改变量值'});
    }
})
```

修改 index.wxml 文件，代码如下。

```
<view>{{arg}}</view>
<input value="{{arg}}"/>
<button bindtap="but_click">点击 button</button>
```

修改 index.wxss 文件，代码如下。

```
input{
  border:1px solid black;
}
input,view,button{
  margin: 100rpx;
}
```

代码解析如下。

- index.js 文件中定义了 arg 变量和 but_click 函数。but_click 函数的作用是修改 arg 变量的值。注意，修改变量值需要使用 this.setData 函数。
- index.wxml 文件中定义的 view 组件用于绑定变量 arg 的值，当 arg 的值修改时，会同步显示。
- index.wxss 文件用于定义样式。

单击“编译”按钮，模拟器中的实现效果如图 3-2 所示。此时，view 显示的是变量 arg 的默认值“this is test”，组件 text 显示的值也是“this is test”。

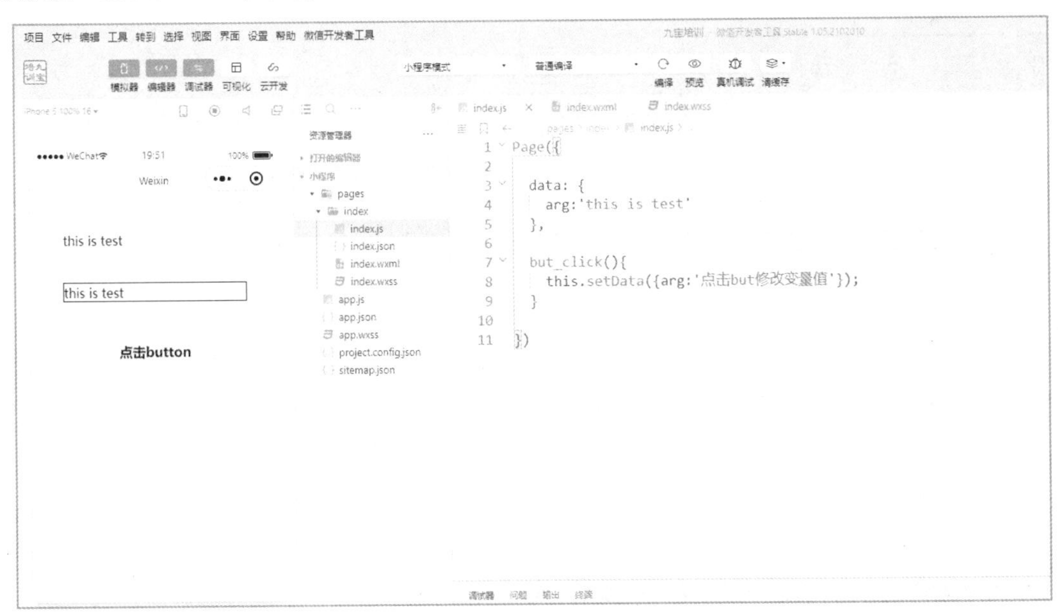

图 3-2

在模拟器中修改 text 的值为“海枯石烂，我心不变”，此时变量 arg 的值保持不变，如图 3-3 所示。

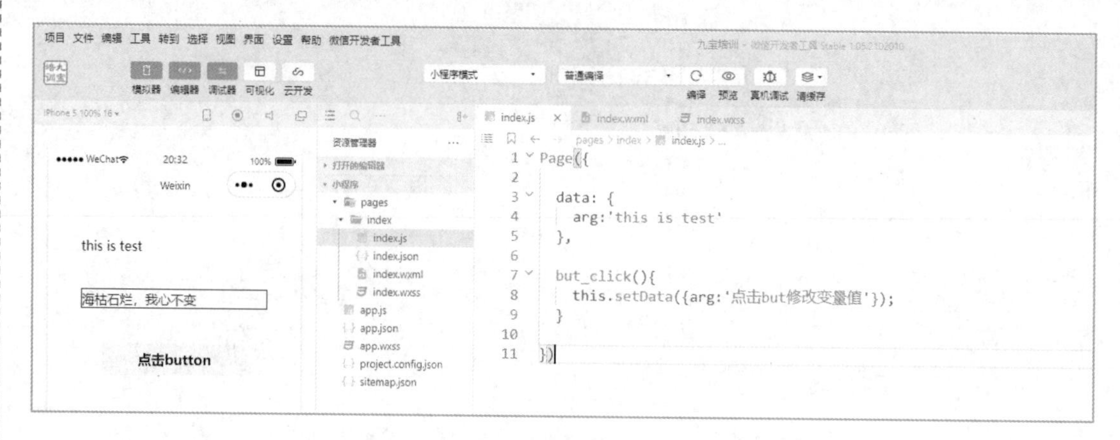

图 3-3

单击“点击 button”按钮，arg 的值会发生改变，view 的显示效果也会发生改变，input 的值同步发生改变，如图 3-4 所示。

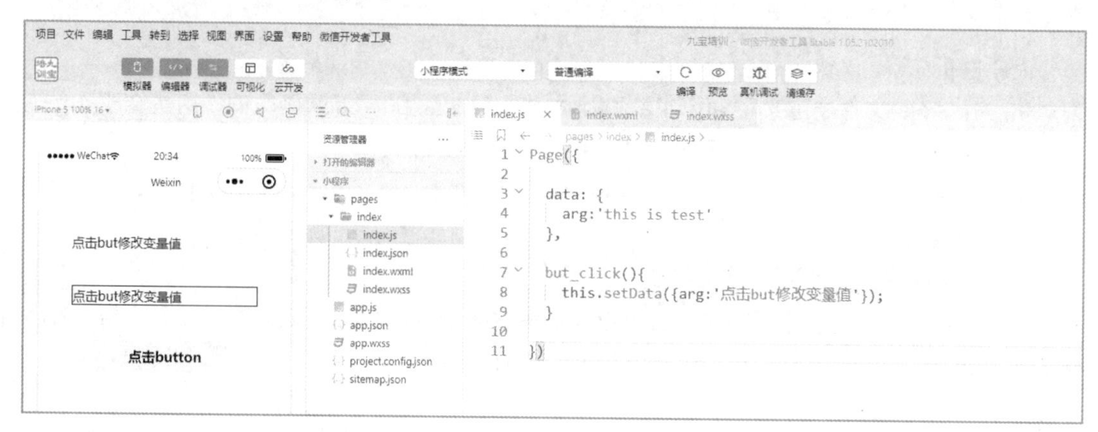

图 3-4

由此说明，微信小程序中的数据绑定是单向数据绑定，这种响应方式在腾讯官方文档中被称为“响应式数据绑定”。

3.4 逻辑层

在微信小程序的逻辑层中，App 方法用来进行程序注册，Page 方法用来进行页面注册，getApp 方法用来获取 App 实例，getCurrentPages 方法用来获取当前页面栈。

注意：

微信小程序框架的逻辑层并不运行在浏览器中。因此，传统 Web 开发中的部分 JavaScript 方法无法使用，如 window、document 等，也不能使用 jQuery、Vue 等框架。

3.4.1 App 方法

构建项目的“最小程序状态”，其中 App 方法的标准定义如下。

```
App({
  /**
   * 当小程序初始化完成时，会触发 onLaunch（全局只触发一次）
   */
  onLaunch: function () {

  },

  /**
   * 当小程序启动，或从后台进入前台显示时，会触发 onShow
   */
  onShow: function (options) {

  },

  /**
   * 当小程序从前台进入后台时，会触发 onHide
   */
  onHide: function () {

  },

  /**
   * 当小程序发生脚本错误，或者 API 调用失败时，会触发 onError 并带上错误信息
   */
  onError: function (msg) {

  }
})
```

微信小程序是独立的程序，包含多个 page。一个微信小程序中只有一个 App 实例，并且是全部页面共享。可以通过 getApp 方法获取全局唯一的 App 实例。

定义全局变量，参考代码如下，开发页面如图 3-5 所示。

```
globalData:"振兴中华"
```

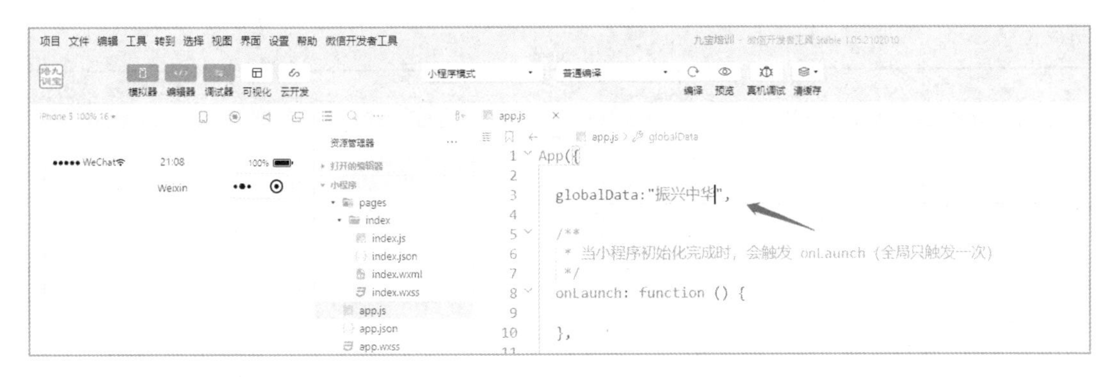

图 3-5

调用全局变量，需要获得 App 实例，参考代码如下，开发页面如图 3-6 所示。

```
getApp().globalData
```

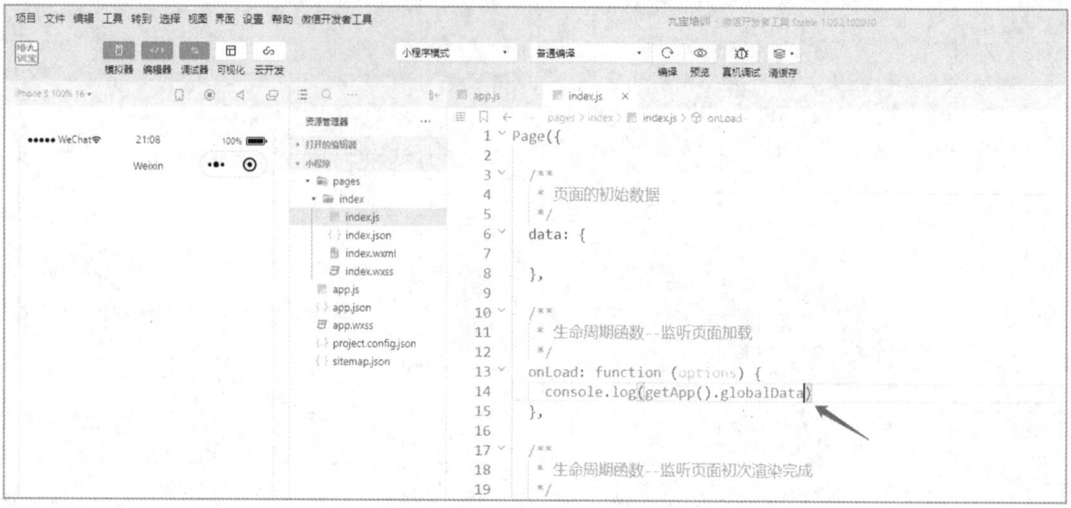

图 3-6

执行编译后，调试器中会打印以下信息。

```
振兴中华
```

App 方法的参数如表 3-1 所示。

表 3-1

属　性	类　型	必　填	说　明	最低版本
onLaunch	function	否	生命周期回调——监听小程序初始化	
onShow	function	否	生命周期回调——监听小程序启动或切换前台	
onHide	function	否	生命周期回调——监听小程序切换后台	
onError	function	否	错误监听函数	
onPageNotFound	function	否	页面不存在监听函数	1.9.90
onUnhandledRejection	function	否	未处理的 Promise 拒绝事件监听函数	2.10.0
onThemeChange	function	否	监听系统主题变化	2.11.0
其他	any	否	开发者可以添加任意的函数或数据变量到 Object 参数中，使用 this 访问	

3.4.2　Page 方法

构建项目的“最小程序状态”，其中 Page 方法的标准定义如下。

```
Page({
  /**
   * 页面的初始数据
   */
  data: {

  },

  /**
   * 生命周期函数--监听页面加载
   */
  onLoad: function (options) {
```

```
  },

  /**
   * 生命周期函数--监听页面初次渲染完成
   */
  onReady: function () {

  },

  /**
   * 生命周期函数--监听页面显示
   */
  onShow: function () {

  },

  /**
   * 生命周期函数--监听页面隐藏
   */
  onHide: function () {

  },

  /**
   * 生命周期函数--监听页面卸载
   */
  onUnload: function () {

  },

  /**
   * 页面相关事件处理函数--监听用户下拉动作
   */
  onPullDownRefresh: function () {

  },

  /**
   * 页面上拉触底事件的处理函数
   */
  onReachBottom: function () {

  },

  /**
   * 用户单击右上角分享
   */
  onShareAppMessage: function () {

  }
})
```

调用自定义方法，参考代码如下。

```
do_some_thing(){
  console.log("微信小程序还能 do_some_thing");
}
```

调用自定义方法需要使用 this，参考代码如下。

```
this.do_some_thing()
```

开发页面如图 3-7 所示。

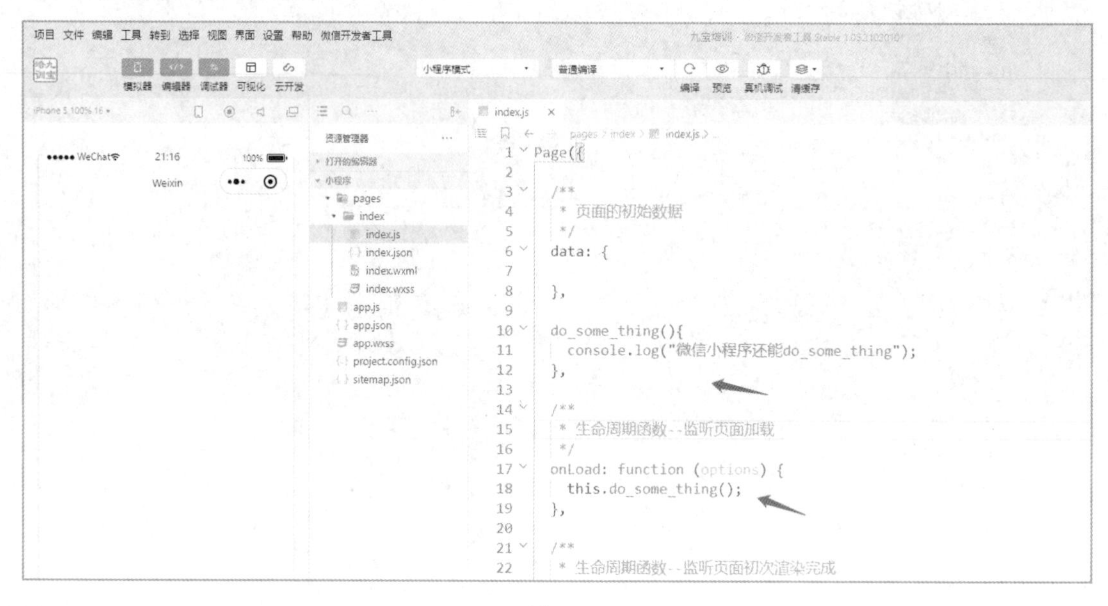

图 3-7

执行编译后，调试器将打印以下信息。

```
微信小程序还能 do_some_thing
```

Page 方法的参数如表 3-2 所示。

表 3-2

属　性	类　型	说　明
data	Object	页面的初始数据
options	Object	页面的组件选项，同 Component 构造器中的 options，需要基础库版本 2.10.1
onLoad	function	生命周期回调——监听页面加载
onShow	function	生命周期回调——监听页面显示
onReady	function	生命周期回调——监听页面初次渲染完成
onHide	function	生命周期回调——监听页面隐藏
onUnload	function	生命周期回调——监听页面卸载
onPullDownRefresh	function	监听用户下拉动作
onReachBottom	function	页面上拉触底事件的处理函数
onShareAppMessage	function	用户单击右上角转发
onShareTimeline	function	用户单击右上角转发到朋友圈
onAddToFavorites	function	用户单击右上角收藏
onPageScroll	function	页面滚动触发事件的处理函数
onResize	function	页面尺寸改变时触发，详见响应显示区域变化
onTabItemTap	function	当前是 tab 页时，单击 tab 时触发
其他	any	开发者可以添加任意的函数或数据到 Object 参数中，在页面的函数中使用 this 访问

3.4.3　getApp 方法

getApp 方法用于获取全局唯一的 App 实例。针对某个 page，可以通过以下代码获取 App 实例。

```
// 获取应用实例
const app = getApp()
```

3.4.4　getCurrentPages 方法

构建项目的“最小程序状态”。page 需要获取当前 this 的需求，可以通过以下代码获取，开发页面如图 3-8 所示。

```
let that = getCurrentPages()[getCurrentPages().length-1];
console.log(this == that);
```

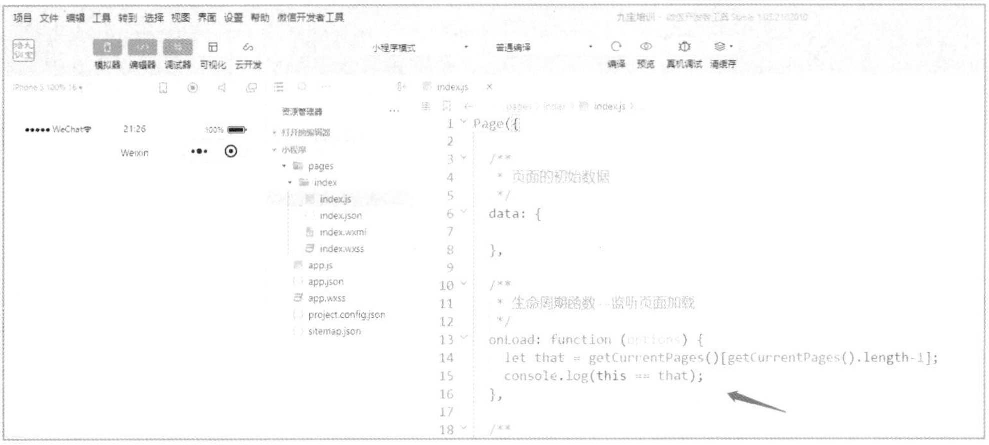

图 3-8

这里，“console.log(this == that);”的作用是判断“getCurrentPages()[getCurrentPages().length-1];”得到的是否是当前 page 的 this。

执行编译后，调试器打印以下信息。

```
true
```

这说明“getCurrentPages()[getCurrentPages().length-1];”得到的是当前 page 的 this。

注意：

大多数微信小程序提供的 API 是异步的，因此在异步返回结果中获取当前 page 的 this 就尤为重要。使用 getCurrentPages 方法是获取当前 page 的解决方案之一。

3.4.5　模块

微信小程序可以将公用程序抽离成一个单独的 JS 文件，使其作为模块使用。模块通过

module.exports 或者 exports 对外暴露接口。exports 是 module.exports 的一个引用，在模块里随意更改 exports 的指向会造成未知的错误。因此，推荐使用 module.exports 来暴露模块接口。

先构建项目的“最小程序状态”，然后在小程序根目录下创建 util/util.js，参考代码如下，开发页面如图 3-9 所示。

```
function speak_chinese() {
  console.log("华语世界")
}

module.exports.speak_chinese = speak_chinese
```

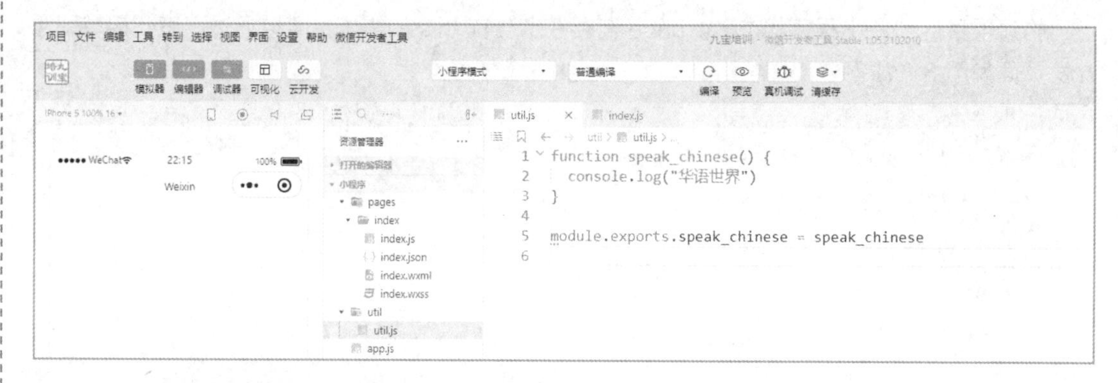

图 3-9

page 可以使用以下代码引用该文件，开发页面如图 3-10 所示。

```
var common = require('../../util/util.js')
common.speak_chinese();
```

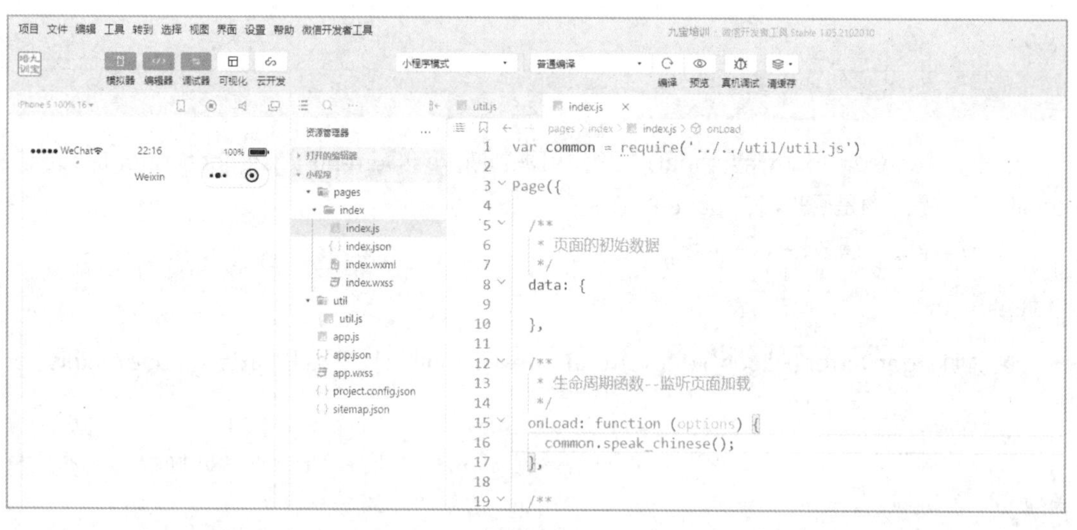

图 3-10

执行编译后，调试器打印以下信息。

```
华语世界
```

3.5　视图层

微信小程序的视图层由 WXML 模板文件与 WXSS 样式文件组成，由组件来进行展示。

其中，WXML（WeiXin Markup Language）模板文件用于描述页面的结构；WXS（WeiXin Script）是小程序的脚本语言，结合 WXML，可以构建页面的结构；WXSS（WeiXin Style Sheet）样式文件用于描述页面的样式；组件（component）是视图的基本组成单元。

3.5.1　WXML

1．数据绑定

WXML 文件的基本功能是进行响应式数据绑定。

构建项目的“最小程序状态”，在 index.js 文件中定义变量，代码参考如下，开发页面如图 3-11 所示。

```
arg:"但使龙城飞将在，不教胡马度阴山"
```

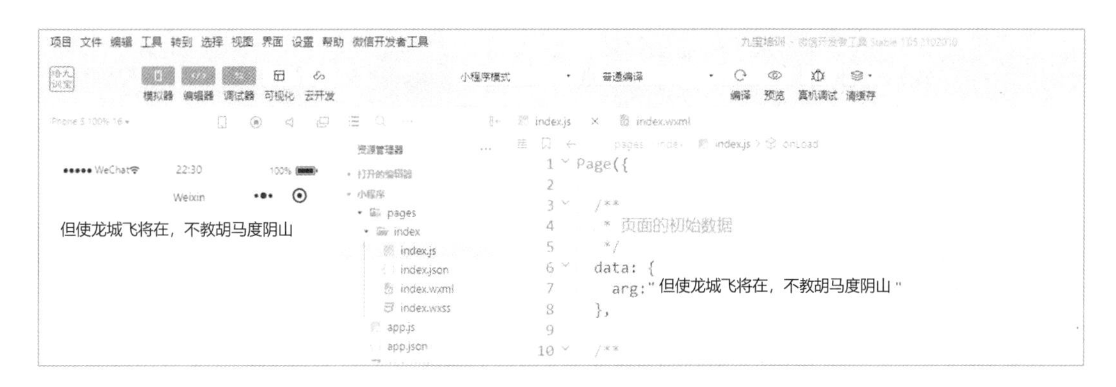

图 3-11

在 index.wxml 文件中绑定变量，参考代码如下，开发页面如图 3-12 所示。

```
{{arg}}
```

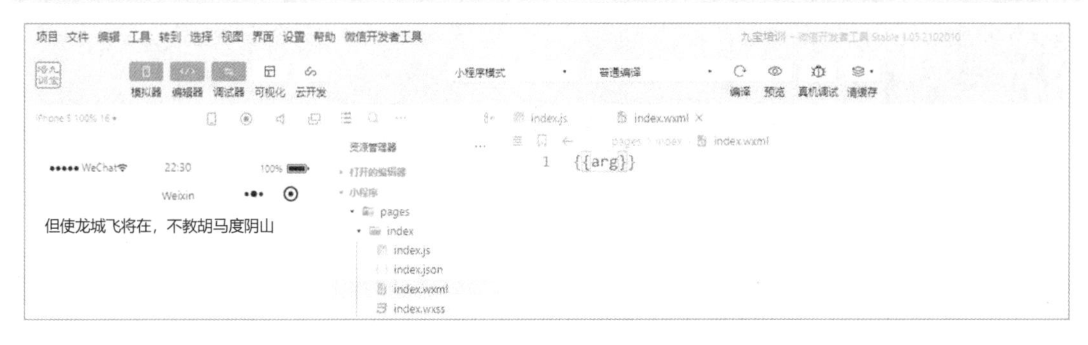

图 3-12

由此便通过{{ }}语法将一个变量绑定到界面上，从而实现数据绑定。

2. 列表渲染

构建项目的“最小程序状态”，在 index.js 文件中定义变量，参考代码如下，开发页面如图 3-13 所示。

```
array:['微信公众号','企业微信','微信小程序','微信支付']
```

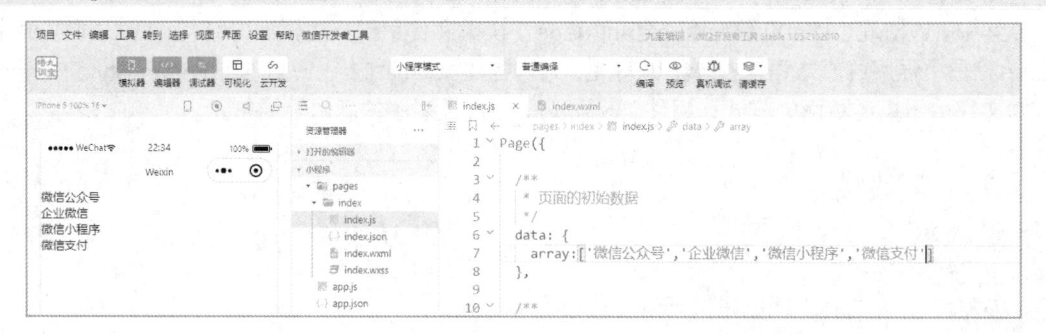

图 3-13

在 index.wxml 文件中绑定变量，参考代码如下，开发页面如图 3-14 所示。

```
<view wx:for="{{array}}"> {{item}} </view>
```

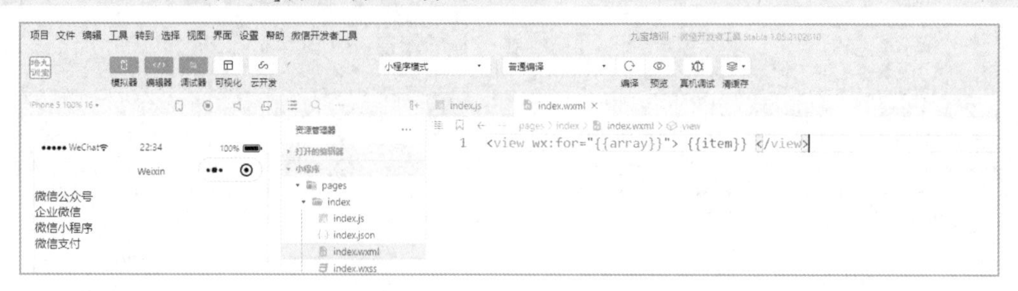

图 3-14

默认情况下，数组当前项的下标变量名为 index，元素变量名为 item。可以使用 wx:for-item 指定数组当前元素的变量名，使用 wx:for-index 指定数组当前下标的变量名。

修改 index.wxml 文件，参考代码如下，开发页面如图 3-15 所示。

```
<view wx:for="{{array}}" wx:for-item="val" wx:for-index="ind">
  {{ind}} - {{val}}
</view>
```

图 3-15

wx:key 用来指定列表中项目的唯一的标识符。

修改 index.js 文件，参考代码如下，开发页面如图 3-16 所示。

```
    data: {
        array:[
            {key:'gzh',val:'微信公众号'},
            {key:'qywx',val:'企业微信'},
            {key:'xcx',val:'微信小程序'},
            {key:'wxpay',val:'微信支付'}
        ]
    },
    but_sort(){
        this.data.array.sort(function(arg,arg_){
            if(arg.key < arg_.key){
              return -1;
            }else{
              return 1;
            }
        });
        this.setData({array:this.data.array});
    },
```

图 3-16

修改 index.wxml 文件，参考代码如下，开发页面如图 3-17 所示。

```
<radio wx:for="{{array}}" wx:for-item="val" wx:for-index="ind">
    {{ind}} - {{val.key}} - {{val.val}}
</radio>

<button bindtap="but_sort">排序</button>
```

图 3-17

执行编译后，先在模拟器中选择“微信小程序”选项，如图 3-18 所示，然后单击“排序”按钮，修改数组排序，结果如图 3-19 所示，可以看到并未实现预期的排序结果。

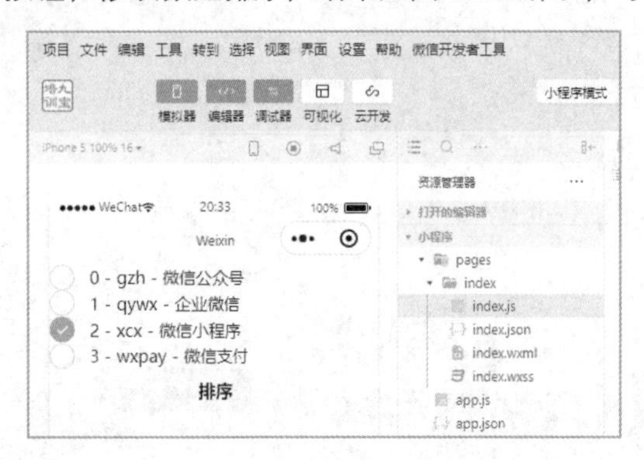

图 3-18

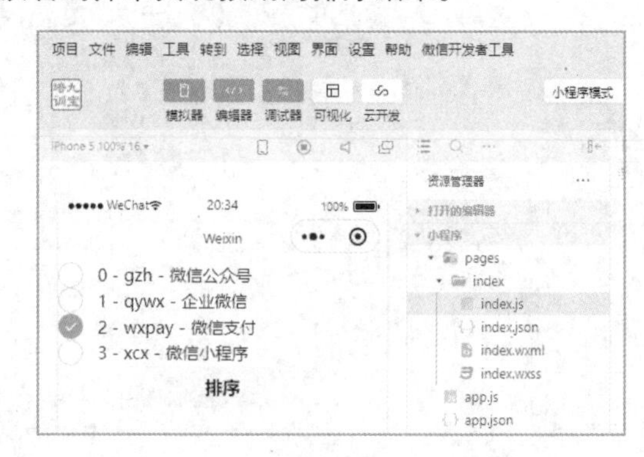

图 3-19

这里存在漏洞，解决这个问题需要使用 wx:key。再次修改 index.wxml 文件，参考代码如下，开发页面如图 3-20 所示。

```
<radio wx:for="{{array}}" wx:for-item="val" wx:for-index="ind" wx:key="key">
  {{ind}} - {{val.key}} - {{val.val}}
</radio>

<button bindtap="but_sort">排序</button>
```

图 3-20

执行编译后，再次在模拟器中选择“微信小程序”选项，如图 3-21 所示。单击“排序”按钮，修改数组排序，此时排序结果如图 3-22 所示，可以看到已经实现了预期的排序效果。

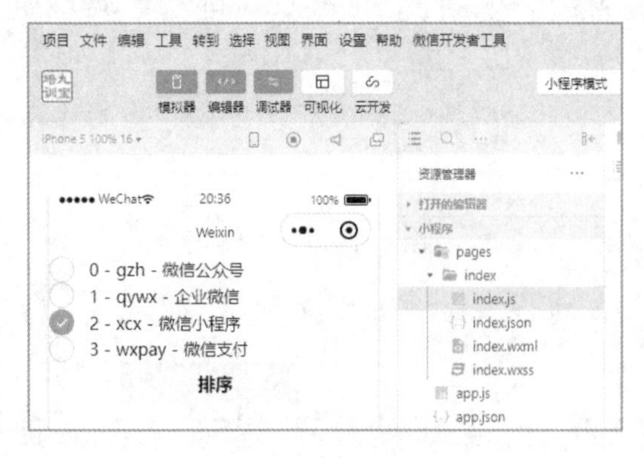

图 3-21

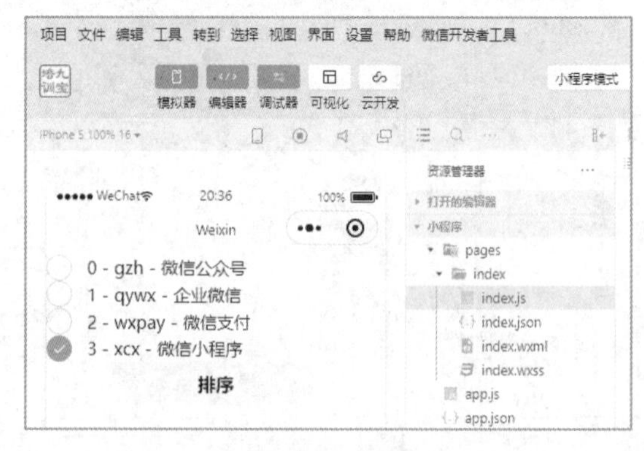

图 3-22

对于静态列表，建议定义 wx:key="*this"，否则会给出 warning 提示。

3.5.2　条件渲染

构建项目的“最小程序状态”，在 index.js 文件中定义变量，参考代码如下，开发页面如图 3-23 所示。

```
wx:'小程序'
```

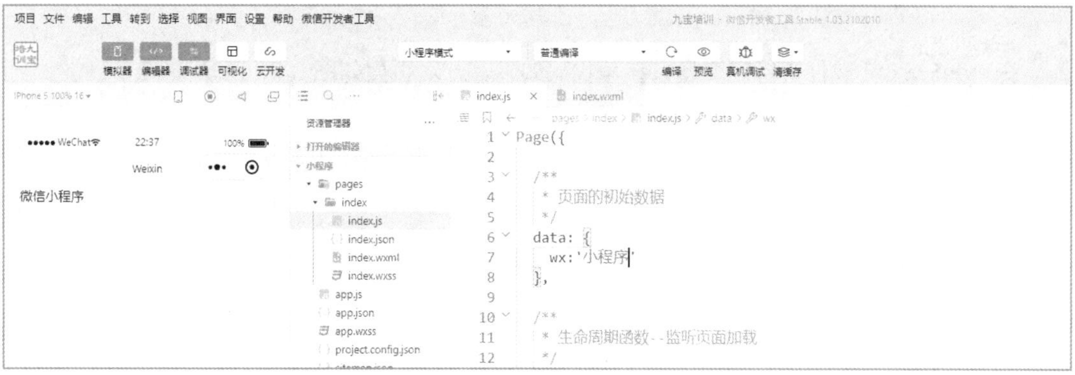

图 3-23

在 index.wxml 文件中绑定变量，参考代码如下，开发页面如图 3-24 所示。

```
<view wx:if="{{wx == '公众号'}}">订阅号、服务号</view>
<view wx:elif="{{wx == '企业微信'}}">企业微信</view>
<view wx:else="{{wx == '小程序'}}">微信小程序</view>
```

图 3-24

3.5.3　模板

构建项目的“最小程序状态”，在 index.js 文件中定义变量，参考代码如下，开发页面如图 3-25 所示。

```
arg:'liubei'
```

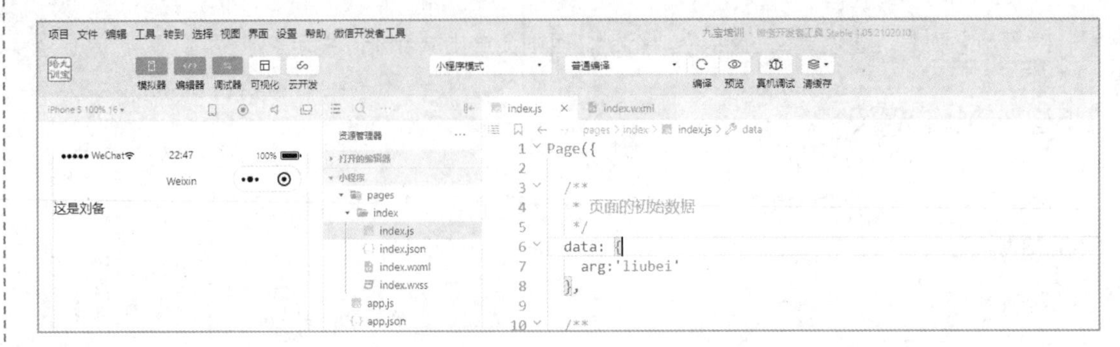

图 3-25

在 index.wxml 文件中绑定变量，参考代码如下，开发页面如图 3-26 所示。

```
<template is="{{arg}}"/>

<template name="liubei">
   <view>
      这是刘备
   </view>
</template>

<template name="guanyu">
   <view>
      这是关羽
   </view>
</template>

<template name="zhangfei">
   <view>
      这是张飞
   </view>
</template>
```

图 3-26

对于需要传递参数的模板，可以使用以下方式。

在 index.js 文件中定义变量，参考代码如下，开发页面如图 3-27 所示。

```
    arg:'liubei',
    wq:'双剑'
```

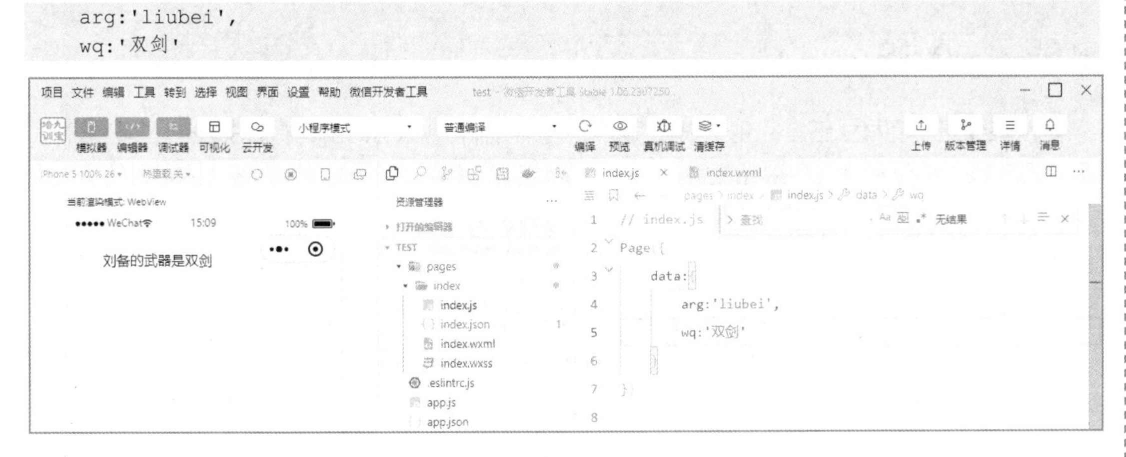

图 3-27

在 index.wxml 文件中绑定变量，参考代码如下，开发页面如图 3-28 所示。

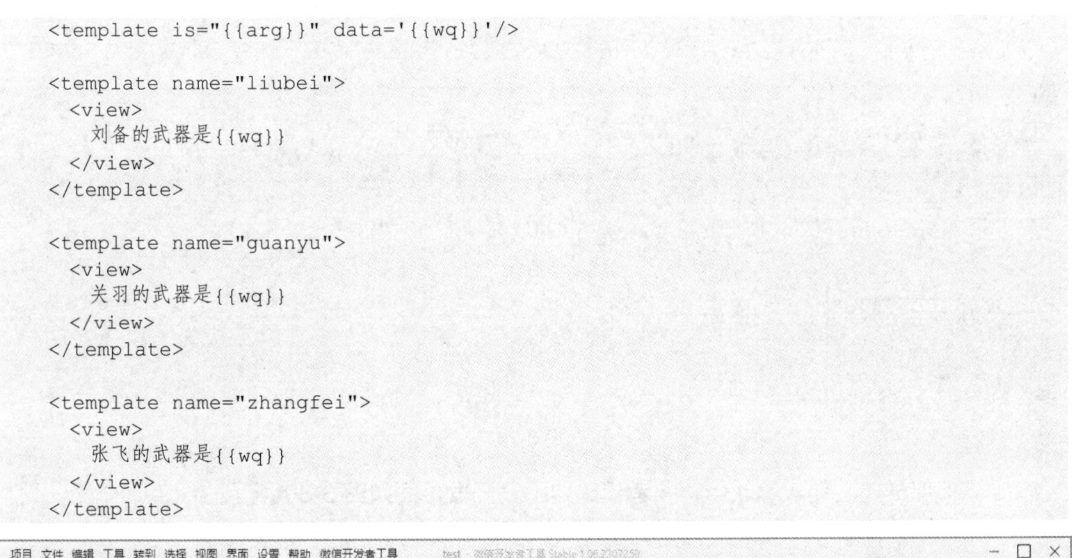

```
<template is="{{arg}}" data='{{wq}}'/>

<template name="liubei">
  <view>
    刘备的武器是{{wq}}
  </view>
</template>

<template name="guanyu">
  <view>
    关羽的武器是{{wq}}
  </view>
</template>

<template name="zhangfei">
  <view>
    张飞的武器是{{wq}}
  </view>
</template>
```

图 3-28

定义模板时，使用 name 属性作为模板的名字。使用模板时，先使用 is 属性声明所需使用的模板，然后将模板所需的 data 传入。

3.5.4 WXSS

WXSS 是一套样式语言，用于描述 WXML 组件的样式。WXSS 具有 CSS 的大部分特性，但扩展了尺寸单位和样式导入特性。

1. 尺寸单位

微信小程序中使用自定义单位 rpx（responsive pixel），可以根据屏幕宽度进行自适应，进而解决了不同移动终端的分辨率问题，这在开发布局小程序时十分有用。

微信小程序规定的屏幕宽度为 750rpx。对于 iPhone 6，其屏幕宽度为 375px，共 750 个物理像素，因此 750rpx = 375px = 750 个物理像素，即 1rpx = 0.5px = 1 个物理像素。

建议读者在开发微信小程序视图层时使用 rpx 来定义组件尺寸。

2. 样式导入

构建项目的“最小程序状态”，在 index.wxml 文件中编写代码，开发页面如图 3-29 所示。

```
<view>这是 view</view>
```

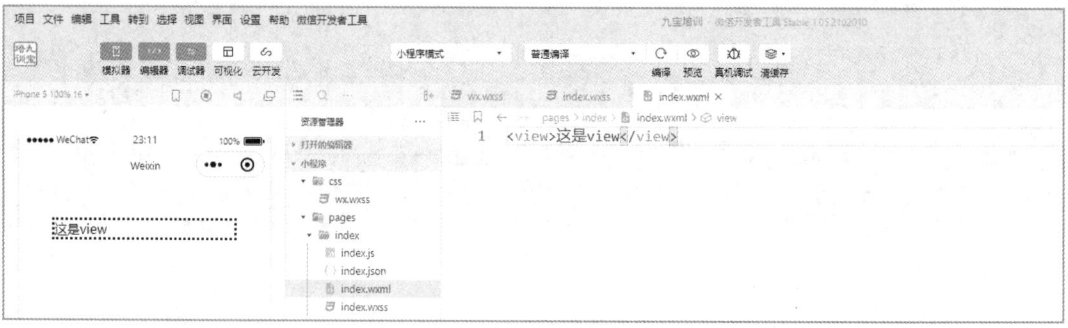

图 3-29

定义外部样式文件 wx.wxss，参考代码如下，开发页面如图 3-30 所示。

```
view{
  border:1px solid black;
}
```

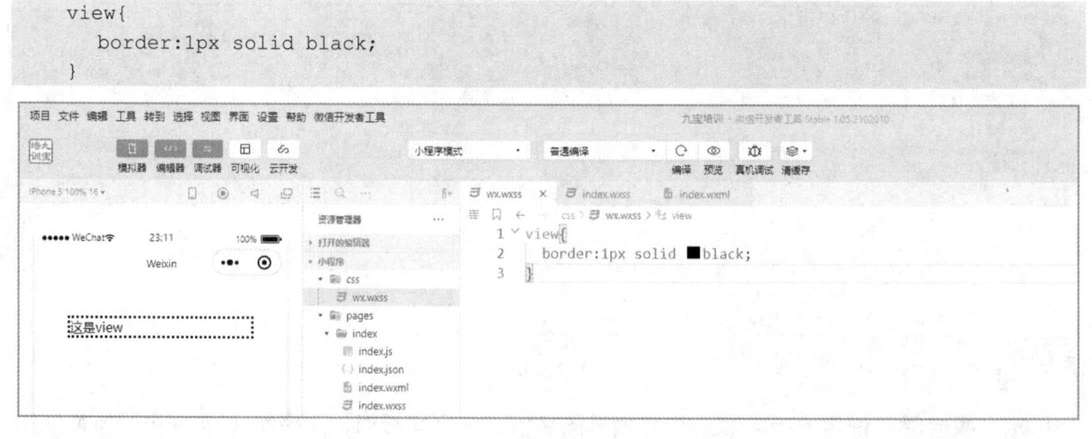

图 3-30

在 index.wxss 文件中导入外部样式，同时定义自己的样式，参考代码如下，开发页面如图 3-31 所示。

```
@import "../../css/wx.wxss";
view{
   margin: 100rpx;
   border: dotted;
}
```

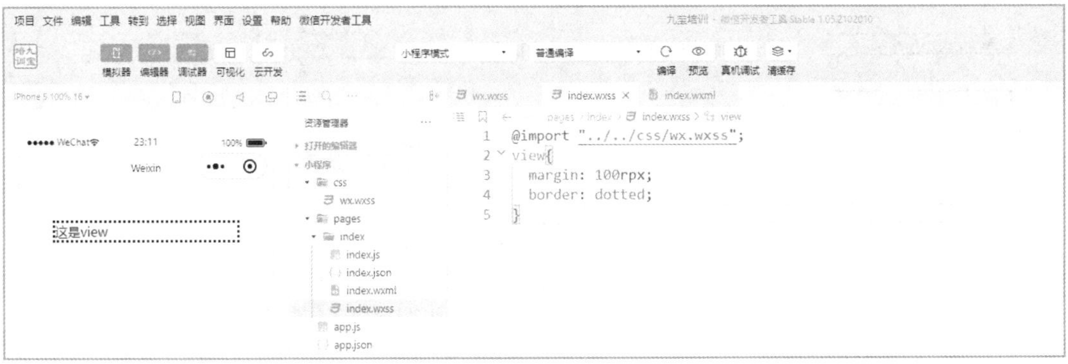

图 3-31

从模拟器中可以看到外部样式与 WXSS 文件同时发挥作用，当定义冲突时，以 WXSS 文件定义覆盖外部样式。

3.5.5　内联样式

构建项目的“最小程序状态”，在 index.js 文件中定义变量，参考代码如下，开发页面如图 3-32 所示。

```
arg:'arg_class'
```

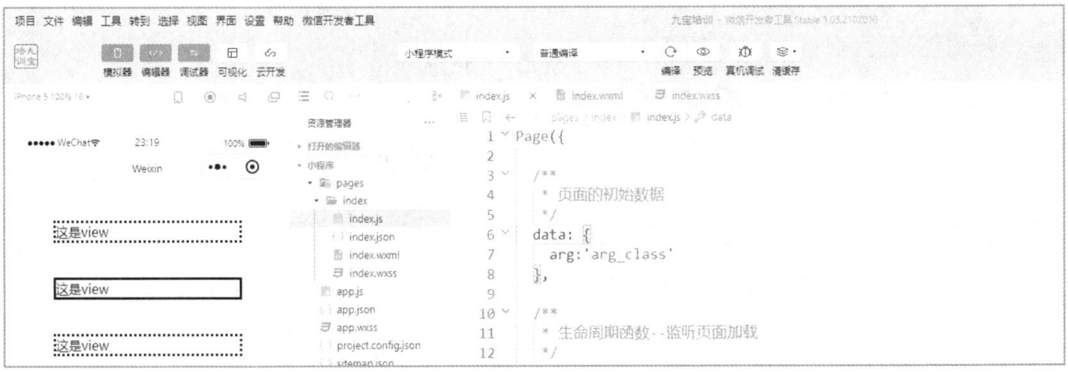

图 3-32

在 index.wxml 文件中绑定变量、定义样式，参考代码如下，开发页面如图 3-33 所示。

```
<view class="class_style">这是 view</view>
<view style="margin:100rpx;  border: solid;">这是 view</view>
<view class="{{arg}}">这是 view</view>
```

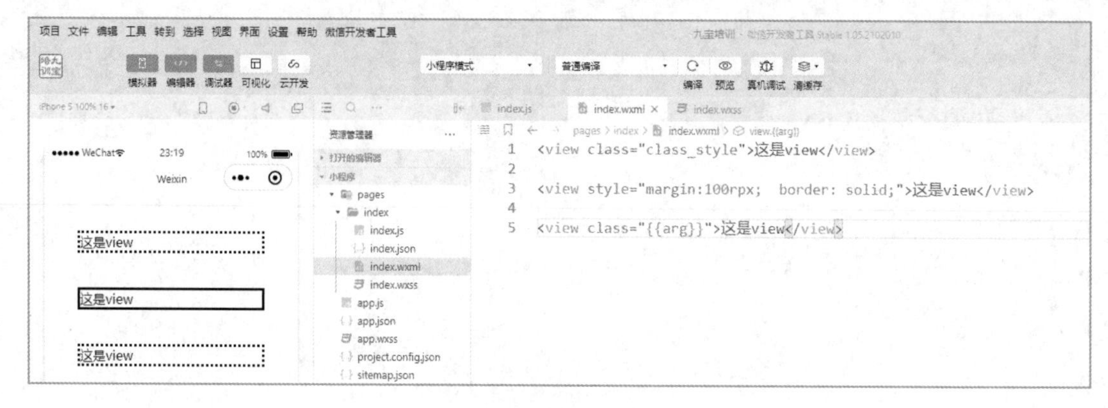

图 3-33

在 index.wxss 文件中定义样式，参考代码如下，开发页面如图 3-34 所示。

```
.class_style{
  margin: 100rpx;
  border: dotted;
}

.arg_class{
  margin: 100rpx;
  border: dotted;
}
```

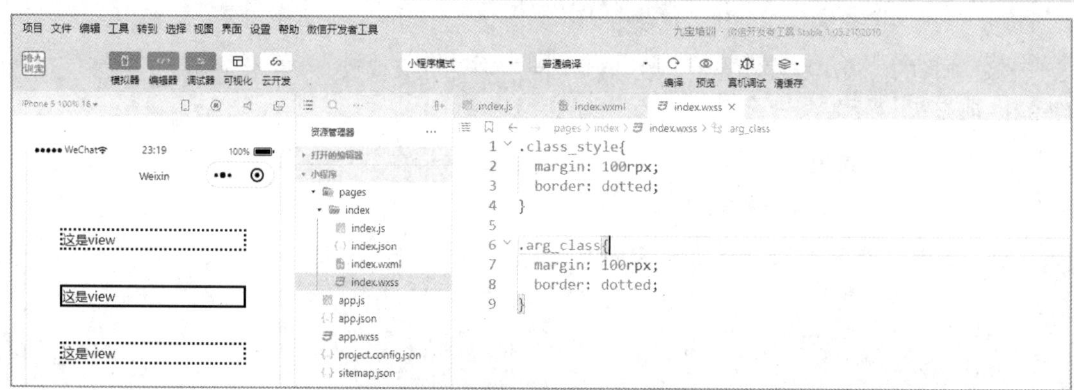

图 3-34

从模拟器中可以看到样式调用都达到了效果。

3.5.6 选择器

微信小程序支持的选择器如表 3-3 所示。

表 3-3

选择器	样例	样例描述
.class	.intro	选择所有拥有 class="intro" 的组件
#id	#firstname	选择拥有 id="firstname" 的组件
element	view	选择所有 view 组件
element, element	view, checkbox	选择所有文档的 view 组件和 checkbox 组件

续表

选　择　器	样　　例	样 例 描 述
::after	view::after	在 view 组件后边插入内容
::before	view::before	在 view 组件前边插入内容

3.6　事件

事件是视图层到逻辑层的通信方式，可以将用户的行为反馈到逻辑层进行处理，同时绑定在组件上，当用户行为触发事件时，就会执行逻辑层中对应的事件处理函数。

3.6.1　简单事件

构建项目的“最小程序状态”，在 index.js 文件中定义函数，参考代码如下，开发页面如图 3-35 所示。

```
but_click(event){
  console.log("这就是按钮事件");
},
```

图 3-35

在 index.wxml 文件中定义按钮组件 button，参考代码如下，开发页面如图 3-36 所示。

```
<button bindtap="but_click">按钮</button>
```

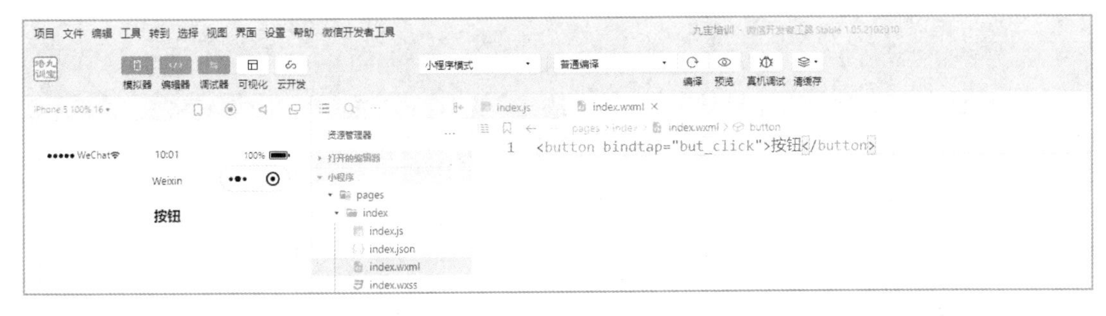

图 3-36

执行编译后，在模拟器中单击“按钮”按钮，调试器将打印以下信息。

```
这就是按钮事件
```

3.6.2 事件参数

组件 button 使用 bindtap 定义事件处理函数 but_click。but_click 函数会接收 event 参数（封装与事件有关的信息）。

在 3.7.1 节中程序的基础上修改函数 but_click，以“console.log(JSON.stringify(event));”方式打印 event 参数，如图 3-37 所示。

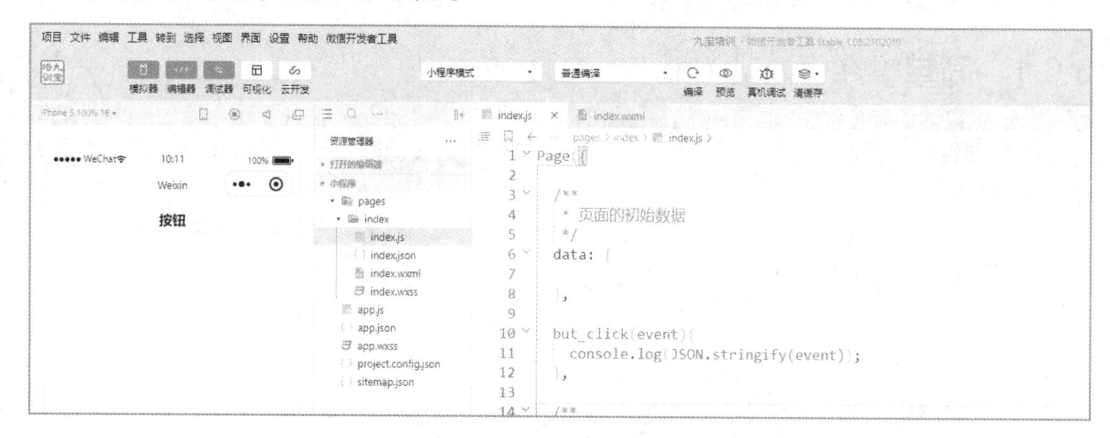

图 3-37

执行编译后，在模拟器中单击“按钮”按钮，将调试器打印信息进行格式化，得到以下报文。

```
{
   "type": "tap",
   "timeStamp": 79937,
   "target": {
      "id": "",
      "offsetLeft": 68,
      "offsetTop": 0,
      "dataset": {}
   },
   "currentTarget": {
      "id": "",
      "offsetLeft": 68,
      "offsetTop": 0,
      "dataset": {}
   },
   "mark": {},
   "detail": {
      "x": 190,
      "y": 22.75
},
   "touches": [{
      "identifier": 0,
      "pageX": 190,
      "pageY": 22.75,
      "clientX": 190,
      "clientY": 22.75,
      "force": 1
   }],
```

```
    "changedTouches": [{
      "identifier": 0,
      "pageX": 190,
      "pageY": 22.75,
      "clientX": 190,
      "clientY": 22.75,
      "force": 1
    }],
    "mut": false,
    "_userTap": true
}
```

3.6.3　事件传参

继续修改 index.wxml 文件，利用 data-arg 方式定义参数 arg 的值为“这是事件参数”。参考代码如下，开发页面如图 3-38 所示。

```
<button bindtap="but_click" data-arg="这是事件参数">按钮</button>
```

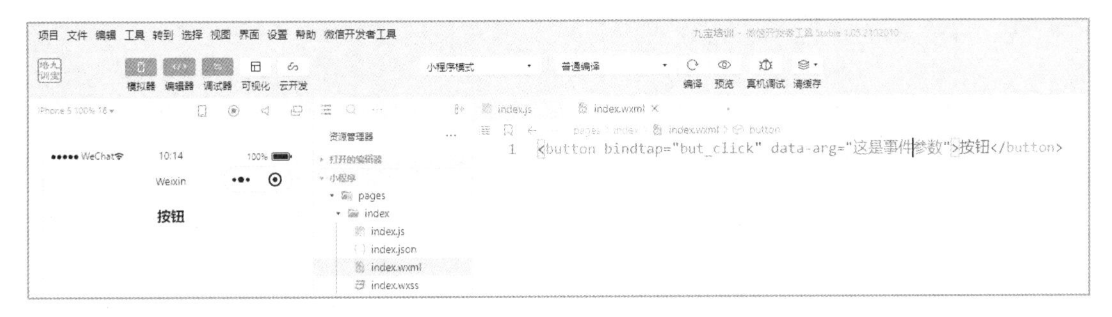

图 3-38

执行编译后，在模拟器中单击“按钮”按钮，将调试器打印信息进行格式化，得到以下报文。

```
{
  "type": "tap",
  "timeStamp": 186021,
  "target": {
    "id": "",
    "offsetLeft": 68,
    "offsetTop": 0,
    "dataset": {
      "arg": "这是事件参数"
    }
  },
  "currentTarget": {
    "id": "",
    "offsetLeft": 68,
    "offsetTop": 0,
    "dataset": {
      "arg": "这是事件参数"
    }
  },
  "mark": {},
  "detail": {
    "x": 181,
    "y": 28.75
  },
  "touches": [{
```

```
        "identifier": 0,
        "pageX": 181,
        "pageY": 28.75,
        "clientX": 181,
        "clientY": 28.75,
        "force": 1
    }],
    "changedTouches": [{
        "identifier": 0,
        "pageX": 181,
        "pageY": 28.75,
        "clientX": 181,
        "clientY": 28.75,
        "force": 1
    }],
    "mut": false,
    "_userTap": true
}
```

特定于这个案例，如需要得到 arg 参数，则可以参考如下代码，开发页面如图 3-39 所示。

```
console.log(event.target.dataset.arg);
console.log(event.currentTarget.dataset.arg);
```

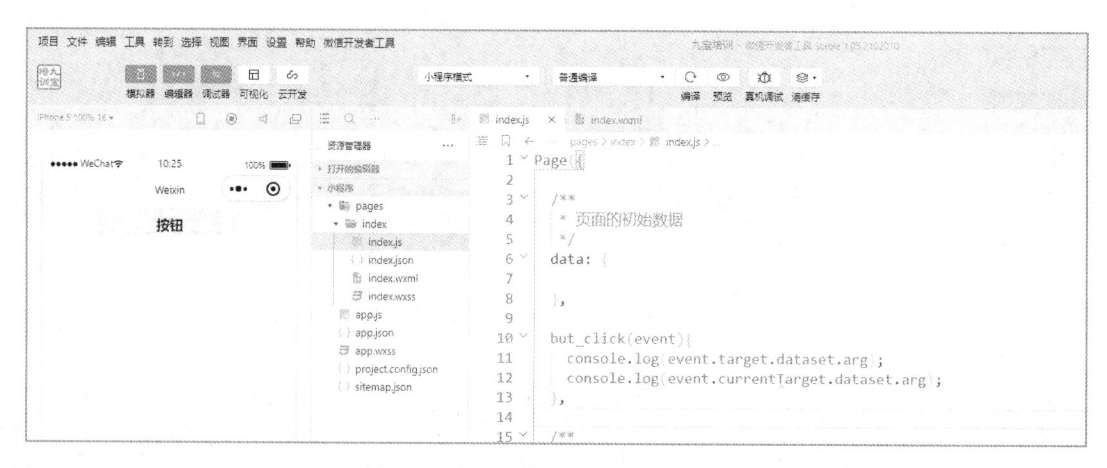

图 3-39

执行编译后，单击“按钮”按钮，调试器将打印以下信息。

```
这是事件参数
这是事件参数
```

3.6.4 事件绑定

微信小程序中的事件可以进行动态绑定。在 3.7.3 节中程序的基础上修改 index.js 文件，参考代码如下，开发页面如图 3-40 所示。

```
data: {
    fun_name:'fun0'
},

fun0(event){
    console.log("这是 fun0");
```

```
},

fun1(event){
    console.log("这是 fun1");
},

change_funName(){
    this.setData({fun_name:'fun1'});
},
```

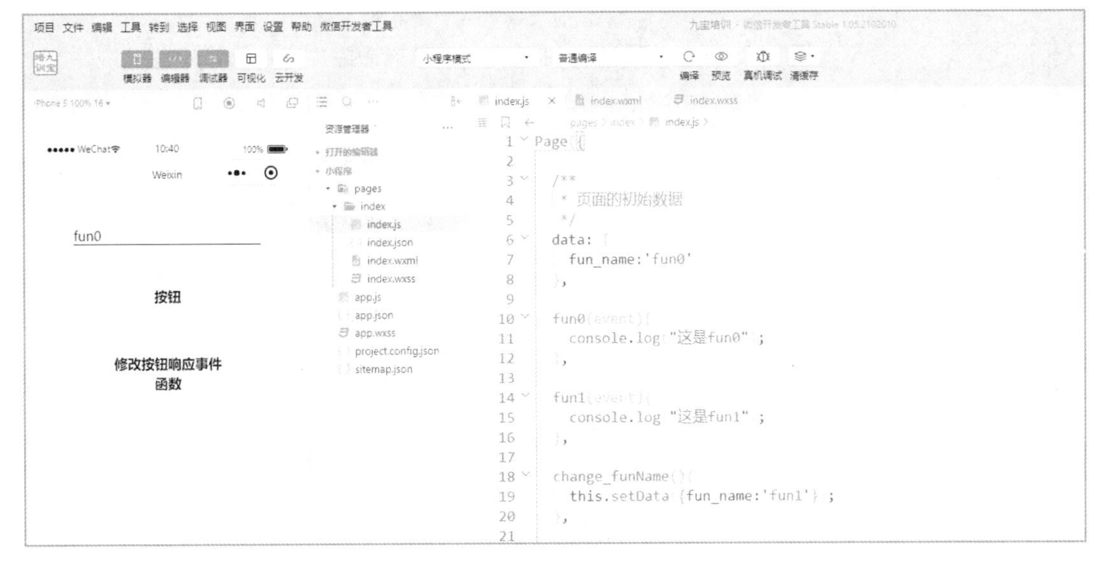

图 3-40

上述程序定义了变量 fun_name，默认值是 fun0，这是按钮默认的响应函数名。

- 函数 fun0 和 fun1 都是用作测试的按钮响应函数。
- 函数 change_funName 是“修改按钮响应事件函数”按钮的事件响应函数，作用是修改变量 fun_name 的值。

修改 index.wxml 文件，参考代码如下，开发页面如图 3-41 所示。

```
<view>{{fun_name}}</view>
<button bindtap="{{fun_name}}">按钮</button>
<button bindtap="change_funName">修改按钮响应事件函数</button>
```

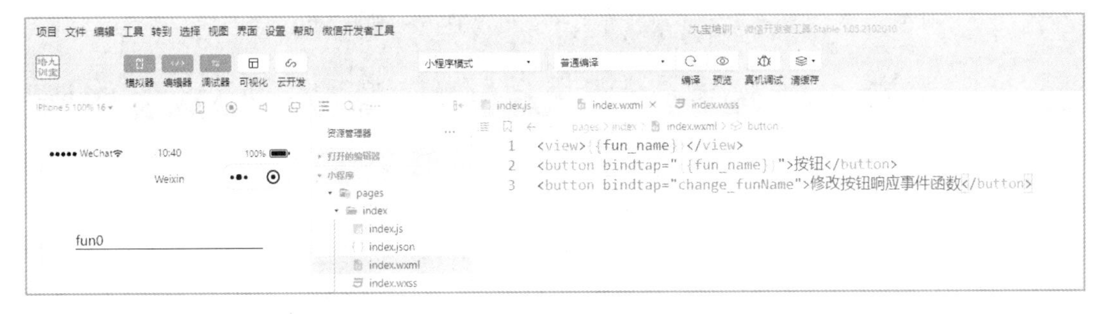

图 3-41

修改 index.wxss 文件，参考代码如下，开发页面如图 3-42 所示。

```
button , view{
```

```
    margin: 100rpx;
}

view{
    border-bottom: 1px solid black;
}
```

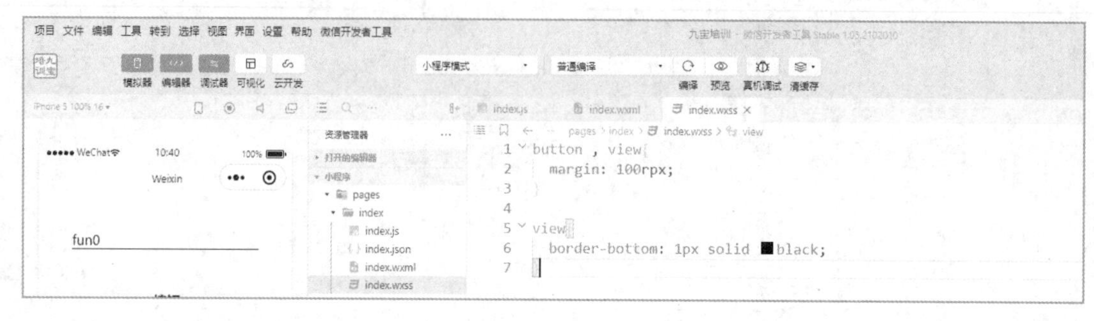

图 3-42

注意：

这里定义的样式只是为了方便测试，与本小节要讲的逻辑知识无必然关系。

执行编译后，模拟器显示效果如图 3-43 所示。

此时，变量 fun_name 的值为 fun0。单击“按钮”按钮，将执行 fun0 响应函数，调试器打印以下信息。

```
这是 fun0
```

先清空调试器，然后单击“修改按钮响应事件函数”按钮，此时变量 fun_name 的值为 fun1，如图 3-44 所示。

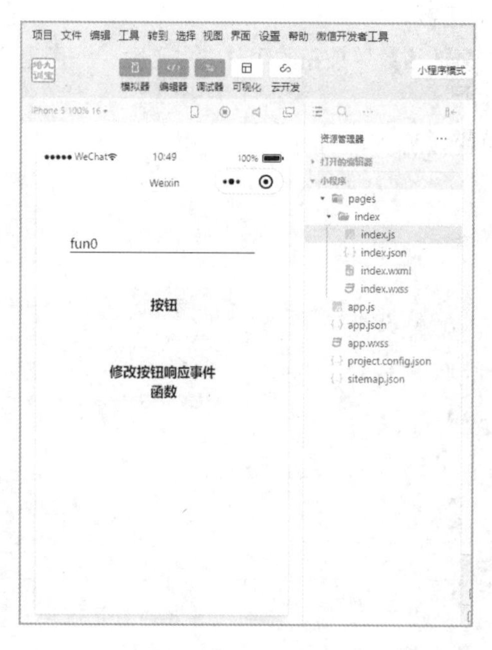

图 3-43

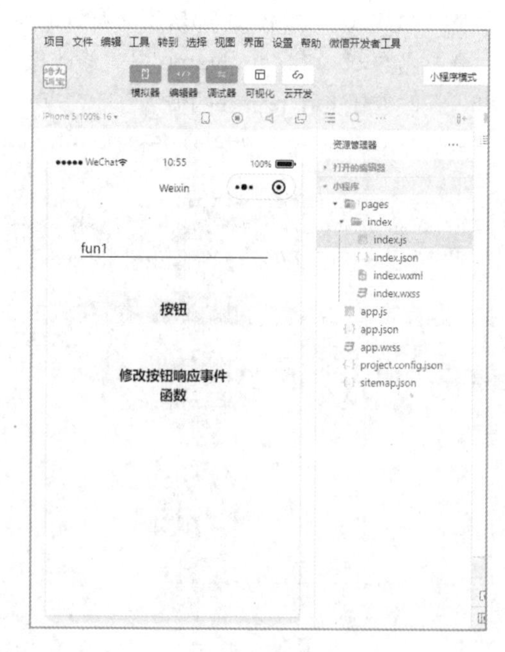

图 3-44

再次单击“按钮”按钮，将执行 fun1 响应函数，调试器打印以下信息。

```
这是 fun1
```

该案例意在讲解微信小程序如何动态绑定事件响应函数。

修改 index.js 文件，重复执行上述过程，打印响应函数的参数，如图 3-45 所示。

图 3-45

调试器打印的信息如下。

（1）fun0 响应函数。

```
{
"type": "tap",
"timeStamp": 1768,
"target": {
    "id": "",
    "offsetLeft": 68,
    "offsetTop": 106,
    "dataset": {}
},
"currentTarget": {
    "id": "",
    "offsetLeft": 68,
    "offsetTop": 106,
    "dataset": {}
},
"mark": {},
"detail": {
    "x": 167,
    "y": 132.75
},
"touches": [{
    "identifier": 0,
    "pageX": 167,
    "pageY": 132.75,
    "clientX": 167,
    "clientY": 132.75,
    "force": 1
}],
```

```
"changedTouches": [{
    "identifier": 0,
    "pageX": 167,
    "pageY": 132.75,
    "clientX": 167,
    "clientY": 132.75,
    "force": 1
}],
"mut": false,
"_userTap": true
}
```

（2）fun1 响应函数。

```
{
"type": "tap",
"timeStamp": 4088,
"target": {
    "id": "",
    "offsetLeft": 68,
    "offsetTop": 106,
    "dataset": {}
},
"currentTarget": {
    "id": "",
    "offsetLeft": 68,
    "offsetTop": 106,
    "dataset": {}
},
"mark": {},
"detail": {
    "x": 181,
    "y": 133.75
},
"touches": [{
    "identifier": 0,
    "pageX": 181,
    "pageY": 133.75,
    "clientX": 181,
    "clientY": 133.75,
    "force": 1
}],
"changedTouches": [{
    "identifier": 0,
    "pageX": 181,
    "pageY": 133.75,
    "clientX": 181,
    "clientY": 133.75,
    "force": 1
}],
"mut": false,
"_userTap": true
}
```

3.6.5 事件冒泡

微信小程序事件分为冒泡事件与非冒泡事件。

- 冒泡事件：组件上的事件被触发后，该事件会向上级节点传递。
- 非冒泡事件：组件上的事件被触发后，该事件不会向上级节点传递。

WXML 的冒泡事件如表 3-4 所示。

表 3-4

类　型	触发条件	最低版本
touchstart	手指触摸动作开始	
touchmove	手指触摸后移动	
touchcancel	手指触摸动作被打断，如来电提醒、弹窗	
touchend	手指触摸动作结束	
tap	手指触摸后马上离开	
longpress	手指触摸后，超过 350 ms 再离开，如果指定了事件回调函数并触发了这个事件，tap 事件将不被触发	1.5.0
longtap	手指触摸后，超过 350 ms 再离开（推荐使用 longpress 事件代替）	
transitionend	会在 WXSS transition 或 wx.createAnimation 动画结束后触发	
animationstart	会在一个 WXSS animation 动画开始时触发	
animationiteration	会在一个 WXSS animation 一次迭代结束时触发	
animationend	会在一个 WXSS animation 动画完成时触发	
touchforcechange	在支持 3D Touch 的 iPhone 设备上重按时会触发	1.9.90

注意：

如无特殊声明，除表 3-4 之外的其他组件自定义事件都是非冒泡事件，如 form 组件的 submit 事件、input 组件的 input 事件、scroll-view 组件的 scroll 事件等。

构建项目的“最小程序状态”，在 index.js 文件中定义变量，参考代码如下，开发页面如图 3-46 所示。

```
handleTap1() {
    console.log("最外层事件");
  },
  handleTap2() {
    console.log("中间层事件");
  },
  handleTap3() {
    console.log("最内层事件");
  },
```

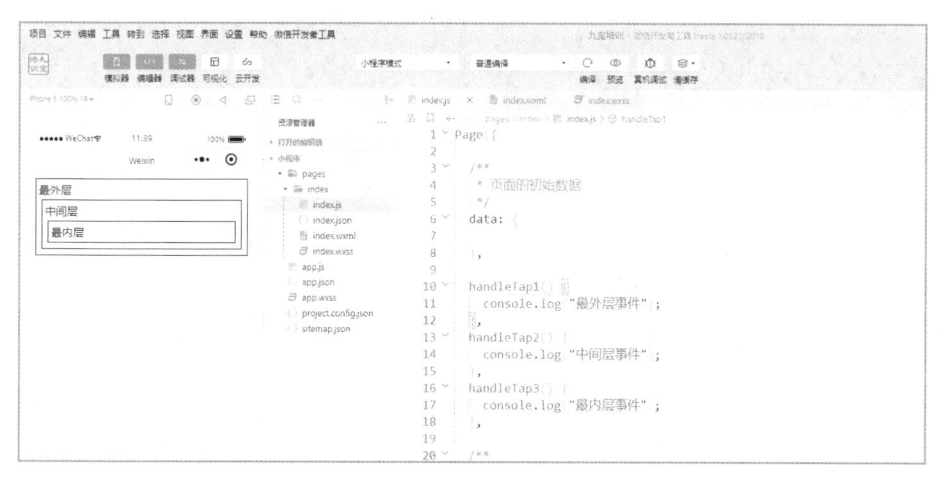

图 3-46

在 index.wxml 文件中绑定变量，参考代码如下，开发页面如图 3-47 所示。

```
<view id="outer" bindtap="handleTap1">
  最外层
  <view id="middle" bindtap="handleTap2">
    中间层
    <view id="inner" bindtap="handleTap3">
      最内层
    </view>
  </view>
</view>
```

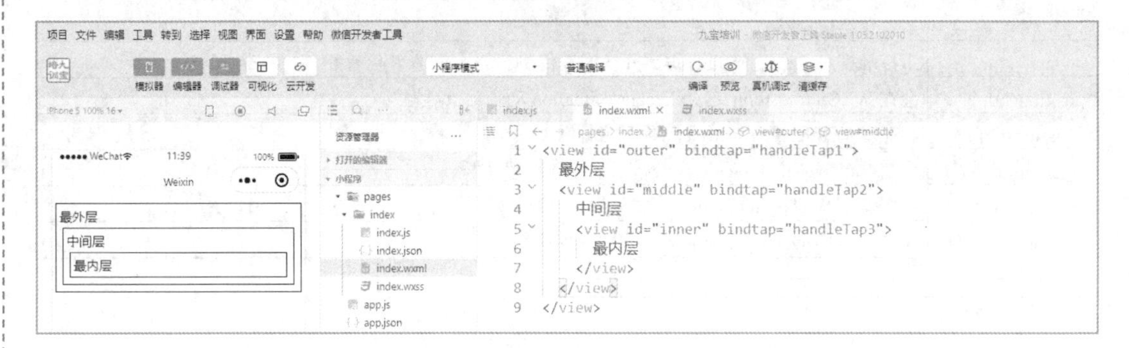

图 3-47

在 index.wxss 文件中绑定变量，参考代码如下，开发页面如图 3-48 所示。

```
view{
  margin: 10rpx;
  padding: 10rpx;
  border: 1px solid black;
}
```

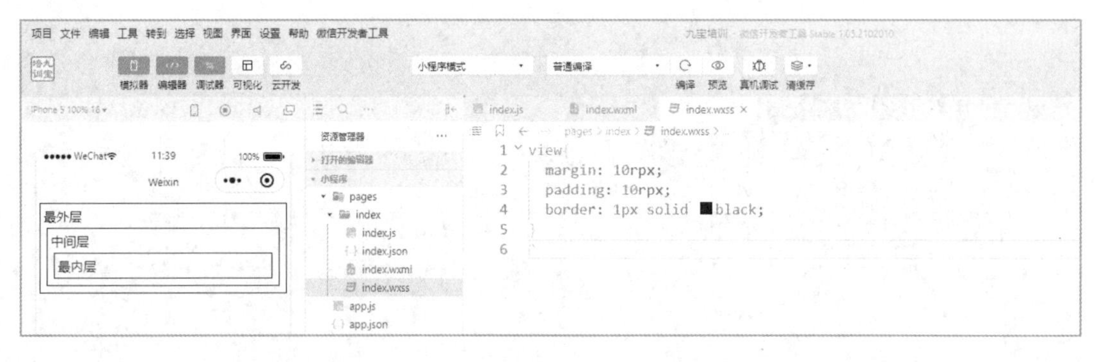

图 3-48

上述程序中，在 index.js 文件中定义的 handleTap1、handleTap2、handleTap3 函数分别用于响应 index.wxml 文件的最外层 view、中间层 view、最内层 view。

handleTap1 函数、handleTap2 函数、handleTap3 函数分别采用 console.log 方式打印信息。

执行编译后，单击“最内层”按钮，调试器打印以下信息，说明最内层事件向上“冒泡”。

```
最内层事件
中间层事件
最外层事件
```

执行编译后，单击“中间层”按钮，调试器打印以下信息，说明中间层事件向上“冒泡”。

```
中间层事件
最外层事件
```

执行编译后，单击“最外层”按钮，调试器打印以下信息，此时只有最外层事件响应函数被执行。

```
最外层事件
```

在此程序基础上修改 index.wxml 文件，参考代码如下，开发页面如图 3-49 所示。

```
<view id="outer" bindtap="handleTap1">
  最外层
  <view id="middle" catchtap="handleTap2">
    中间层
    <view id="inner" bindtap="handleTap3">
      最内层
    </view>
  </view>
</view>
```

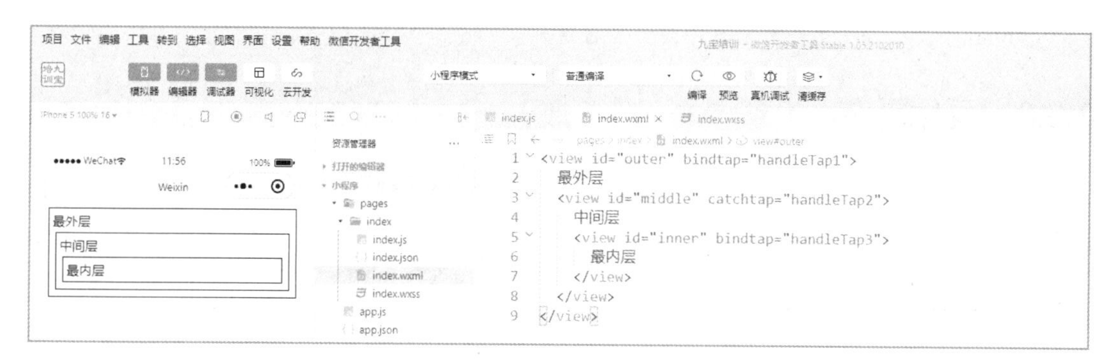

图 3-49

▶ **注意：**

这里将中间层事件方式改为 catchtap。

执行编译后，单击“最内层”按钮，调试器打印以下信息。

```
最内层事件
中间层事件
```

此时，最内层事件被触发，同时事件向上“冒泡”。随后中间层事件被触发，但由于中间层是非冒泡事件，因此该事件不会再次向上传递。

在此程序基础上修改 index.wxml 文件，参考代码如下，开发页面如图 3-50 所示。

```
<view id="outer" bindtap="handleTap1">
  最外层
  <view id="middle" bindtap="handleTap2">
    中间层
    <view id="inner" catchtap="handleTap3">
      最内层
    </view>
  </view>
</view>
```

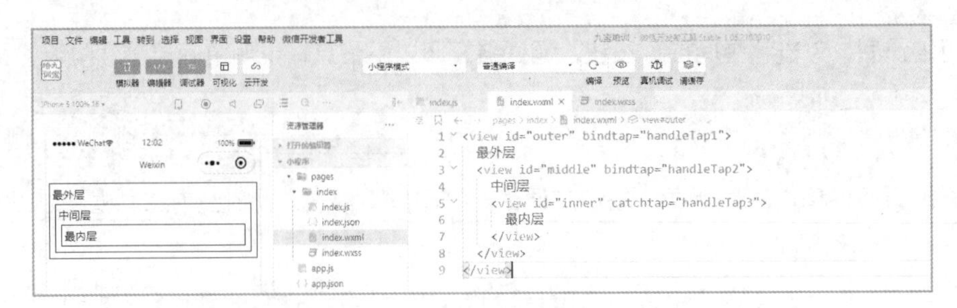

图 3-50

注意：

这里修改的是最内层事件方式。

执行编译后，单击“最内层”按钮，调试器打印以下信息。由于最内层是非冒泡事件，因此只有最内层事件被触发。

```
最内层事件
```

注意：

与 bind 不同， catch 会阻止事件向上“冒泡”。

3.6.6　互斥事件

构建项目的“最小程序状态”，在 index.js 文件中定义变量，参考代码如下，开发页面如图 3-51 所示。

```
handleTap1() {
  console.log("最外层事件");
},
handleTap2() {
  console.log("中间层事件");
},
handleTap3() {
  console.log("最内层事件");
},
```

图 3-51

在 index.wxml 文件中绑定变量，参考代码如下，开发页面如图 3-52、图 3-53 所示。

```
<view id="outer" mut-bind:tap="handleTap1">
  最外层
  <view id="middle" bindtap="handleTap2">
    中间层
    <view id="inner" mut-bind:tap="handleTap3">
      最内层
    </view>
  </view>
</view>

<view id="outer" mut-bind:tap="handleTap1">
  最外层
  <view id="middle" catchtap="handleTap2">
    中间层
    <view id="inner" mut-bind:tap="handleTap3">
      最内层
    </view>
  </view>
</view>

<view id="outer" bindtap="handleTap1">
  最外层
  <view id="middle" mut-bind:tap="handleTap2">
    中间层
    <view id="inner" bindtap="handleTap3">
      最内层
    </view>
  </view>
</view>

<view id="outer" catchtap="handleTap1">
  最外层
  <view id="middle" mut-bind:tap="handleTap2">
    中间层
    <view id="inner" catchtap="handleTap3">
      最内层
    </view>
  </view>
</view>
```

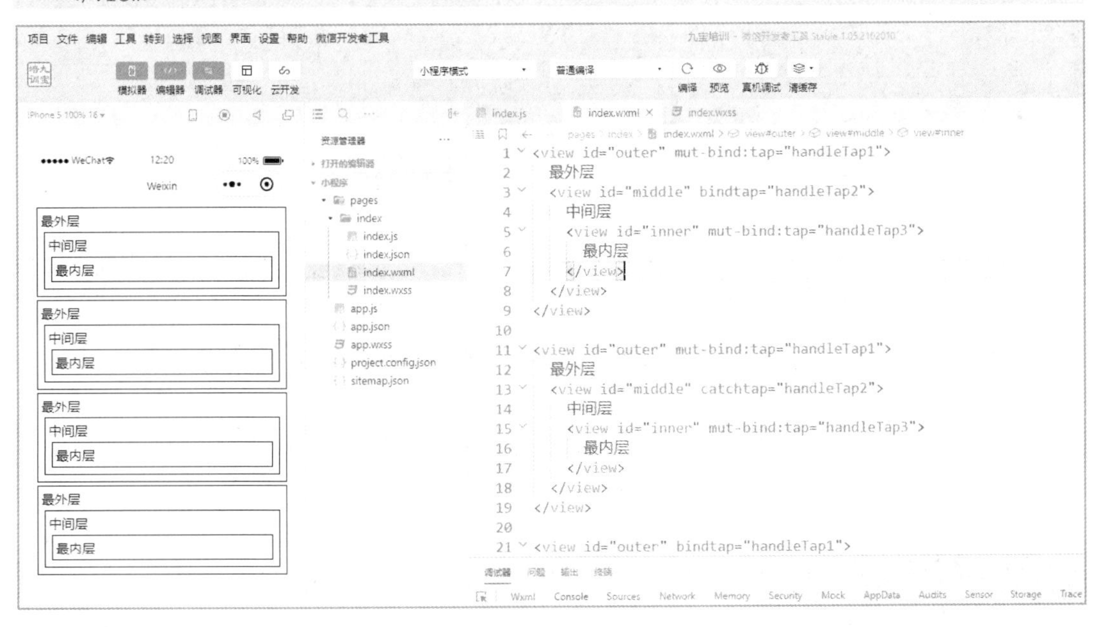

图 3-52

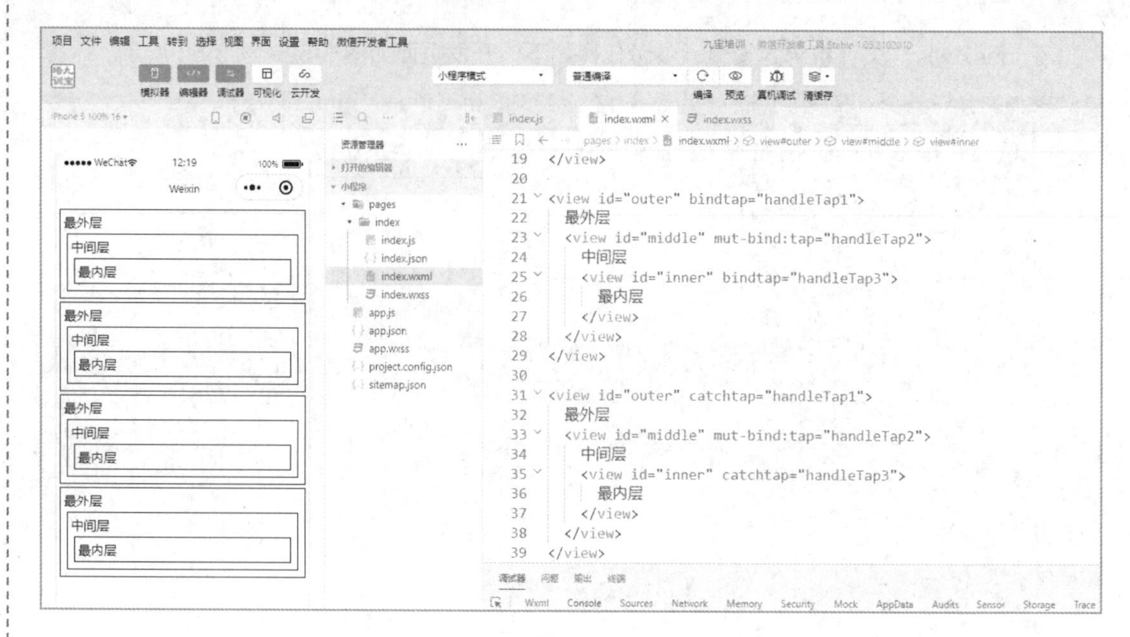

图 3-53

在 index.wxss 文件中绑定变量，参考代码如下，开发页面如图 3-54 所示。

```
view{
  margin: 10rpx;
  padding: 10rpx;
  border: 1px solid black;
}
```

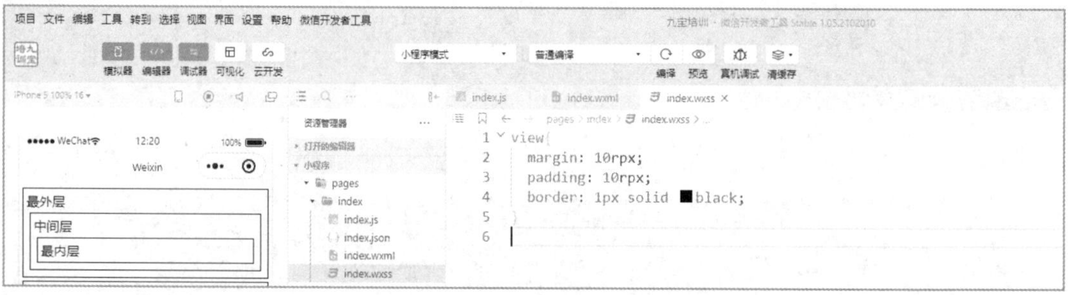

图 3-54

上述程序中，在 index.js 文件中定义的 handleTap1、handleTap2、handleTap3 函数分别用于响应 index.wxml 文件的最外层 view、中间层 view、最内层 view。

handleTap1、handleTap2、handleTap3 函数分别以 console.log 方式打印信息。

需要重点说明的是 WXML 文件。

（1）第一组 view 结构为 mut-bind:tap、bindtap、mut-bind:tap。

单击“最内层”按钮，调试器打印以下信息。

```
最内层事件
中间层事件
```

单击“中间层”按钮，调试器打印以下信息。

```
中间层事件
最外层事件
```

单击“最外层”按钮，调试器打印以下信息。

```
最外层事件
```

（2）第二组 view 结构为 mut-bind:tap、catchtap、mut-bind:tap。

单击“最内层”按钮，调试器打印以下信息。

```
最内层事件
中间层事件
```

单击“中间层”按钮，调试器打印以下信息。

```
中间层事件
```

单击“最外层”按钮，调试器打印以下信息。

```
最外层事件
```

（3）第三组 view 结构为 bindtap、mut-bind:tap、bindtap。

单击“最内层”按钮，调试器打印以下信息。

```
最内层事件
中间层事件
最外层事件
```

单击“中间层”按钮，调试器打印以下信息。

```
中间层事件
最外层事件
```

单击“最外层”按钮，调试器打印以下信息。

```
最外层事件
```

（4）第四组 view 结构为 catchtap、mut-bind:tap、catchtap。

单击“最内层”按钮，调试器打印以下信息。

```
最内层事件
```

单击“中间层”按钮，调试器打印以下信息。

```
中间层事件
最外层事件
```

单击“最外层”按钮，调试器打印以下信息。

```
最外层事件
```

综上，mut-bind 之间是“互斥”的。同时，mut-bind 不影响 bind 和 catch 的绑定效果。

3.6.7　事件的捕获阶段

构建项目的“最小程序状态”。在 index.js 文件中定义变量，参考代码如下，开发页面如

图 3-55 所示。

```
handleTap1() {
    console.log("最外层事件");
  },
  handleTap2() {
    console.log("中间层事件");
  },
  handleTap3() {
    console.log("最内层事件");
  },

  capture_handleTap1() {
    console.log("最外层捕获阶段");
  },
  capture_handleTap2() {
    console.log("中间层捕获阶段");
  },
  capture_handleTap3() {
    console.log("最内层捕获阶段");
  },
```

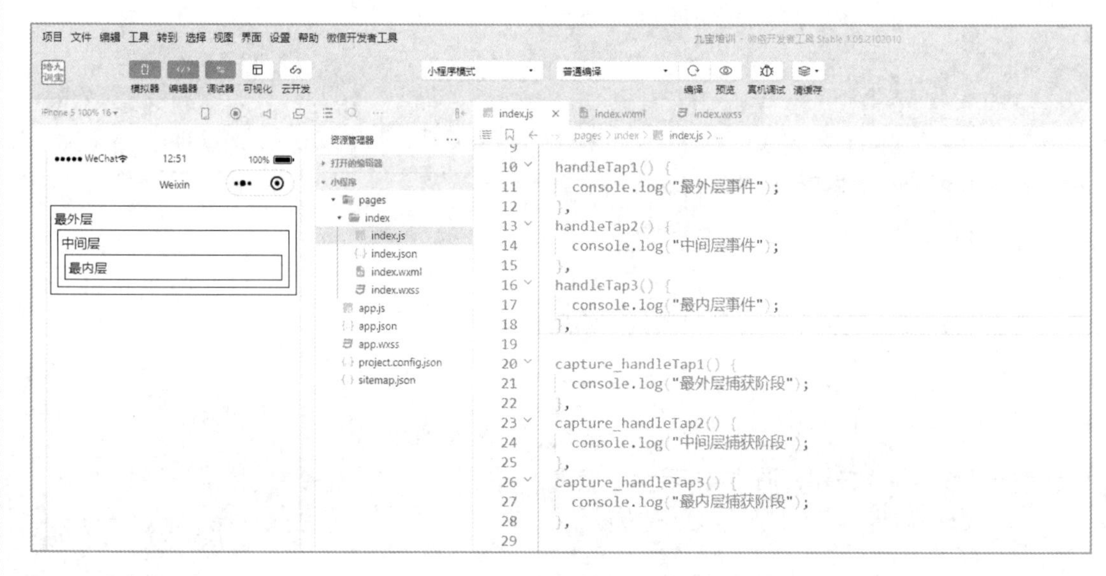

图 3-55

在 index.wxml 文件中绑定变量，参考代码如下，开发页面如图 3-56 所示。

```
<view id="outer" bind:tap="handleTap1" capture-bind:tap="capture_handleTap1">
  最外层
  <view id="middle" bindtap="handleTap2" capture-bind:tap="capture_handleTap2">
    中间层
    <view id="inner" bind:tap="handleTap3" capture-bind:tap="capture_handleTap3">
      最内层
    </view>
  </view>
</view>
```

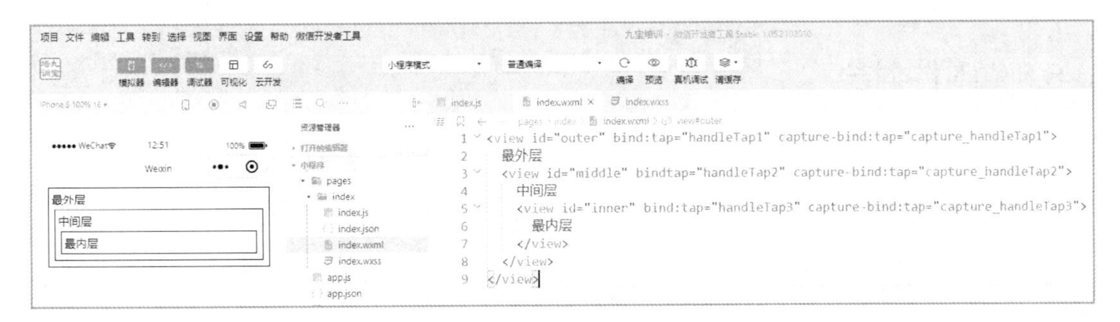

图 3-56

在 index.wxss 文件中绑定变量，参考代码如下，开发页面如图 3-57 所示。

```
view{
  margin: 10rpx;
  padding: 10rpx;
  border: 1px solid black;
}
```

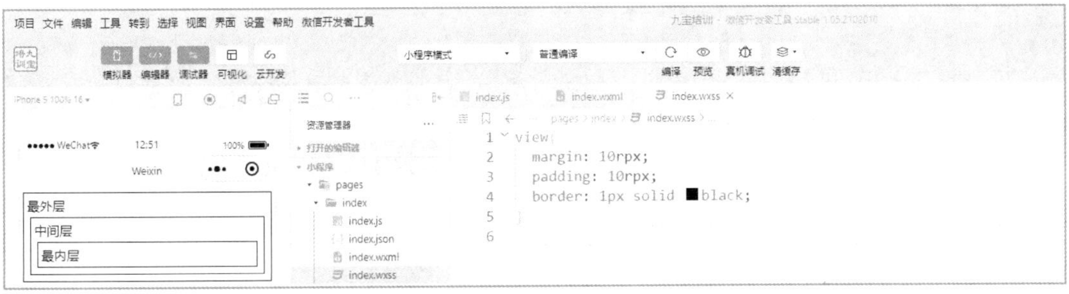

图 3-57

单击“最内层”按钮，调试器打印以下信息。

```
最外层捕获阶段
中间层捕获阶段
最内层捕获阶段
最内层事件
中间层事件
最外层事件
```

单击“中间层”按钮，调试器打印以下信息。

```
最外层捕获阶段
中间层捕获阶段
中间层事件
最外层事件
```

单击“最外层”按钮，调试器打印以下信息。

```
最外层捕获阶段
最外层事件
```

捕获阶段会先于事件冒泡执行，且在捕获阶段中事件捕获节点的顺序与冒泡阶段恰好相反。

3.6.8 事件对象

构建项目的“最小程序状态”，在 index.js 文件中定义函数，参考代码如下，开发页面如图 3-58 所示。

```
but_click(event){
  console.log(JSON.stringify(event));
},
```

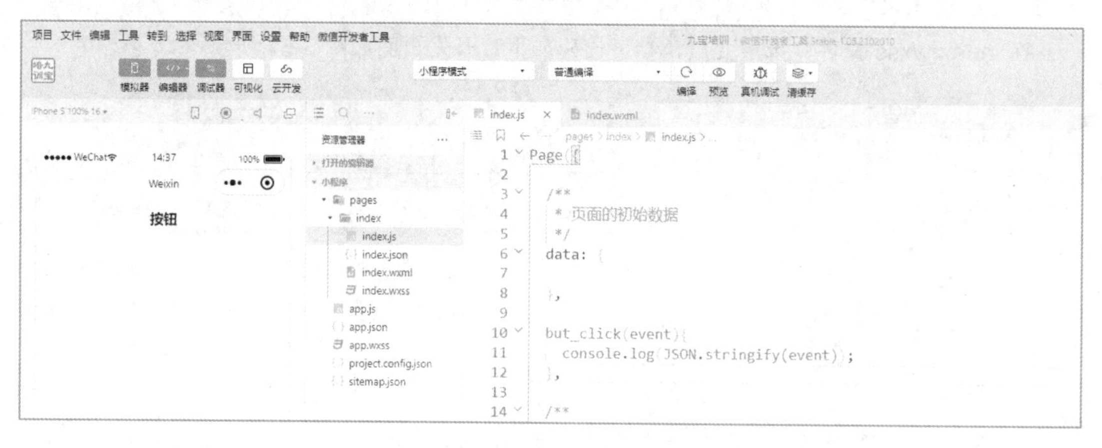

图 3-58

在 index.wxml 文件中绑定函数，参考代码如下，开发页面如图 3-59 所示。

```
<button bindtap="but_click">按钮</button>
```

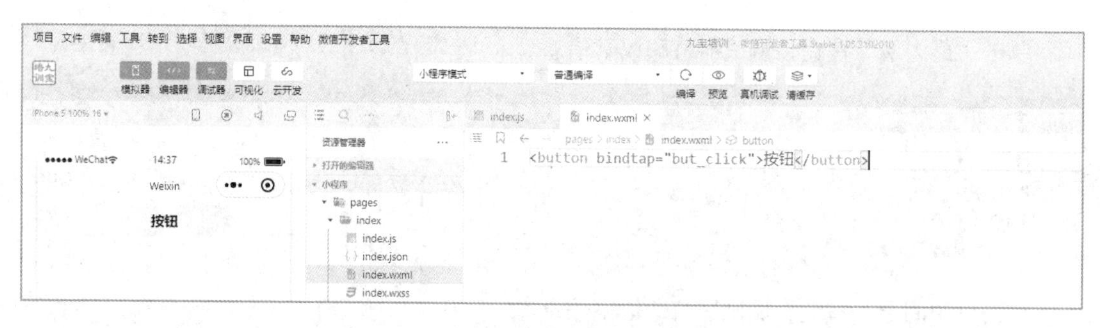

图 3-59

执行编译后，单击“按钮”按钮，格式化调试器打印信息，如下所示。

```
{
"type": "tap",
"timeStamp": 162367,
"target": {
     "id": "",
     "offsetLeft": 68,
     "offsetTop": 0,
     "dataset": {}
},
"currentTarget": {
```

```
        "id": "",
        "offsetLeft": 68,
        "offsetTop": 0,
        "dataset": {}
    },
    "mark": {},
    "detail": {
        "x": 199,
        "y": 19.75
    },
    "touches": [{
        "identifier": 0,
        "pageX": 199,
        "pageY": 19.75,
        "clientX": 199,
        "clientY": 19.75,
        "force": 1
    }],
    "changedTouches": [{
        "identifier": 0,
        "pageX": 199,
        "pageY": 19.75,
        "clientX": 199,
        "clientY": 19.75,
        "force": 1
    }],
    "mut": false,
    "_userTap": true
    }
```

一般当组件触发事件时，逻辑层绑定该事件的处理函数会收到一个事件对象。相关信息参考表 3-5。

表 3-5

类　型	属　性	类　型	说　明	基础库版本
BaseEvent	type	String	事件类型	
	timeStamp	Integer	事件生成时的时间戳	
	target	Object	触发事件的组件的一些属性值集合	
	currentTarget	Object	当前组件的一些属性值集合	
	mark	Object	事件标记数据	2.7.1
CustomEvent	detail	Object	额外的信息	
TouchEvent	touches	Array	触摸事件，当前停留在屏幕中的触摸点信息的数组	
	changedTouches	Array	触摸事件，当前变化的触摸点信息的数组	
target	id	String	事件源组件的 id	
	dataset	Object	事件源组件上由 data-开头的自定义属性组成的集合	
currentTarget	id	String	当前组件的 id	
	dataset	Object	当前组件上由 data-开头的自定义属性组成的集合	

续表

类　型	属　性	类　型	说　明	基础库版本
Touch	identifier	Number	触摸点的标识符	
	pageX, pageY	Number	距离文档左上角的距离，文档左上角为原点，横向为 X 轴，纵向为 Y 轴	
	clientX, clientY	Number	距离页面可显示区域（屏幕除去导航条）左上角的距离，横向为 X 轴，纵向为 Y 轴	
CanvasTouch	identifier	Number	触摸点的标识符	
	x, y	Number	距离 Canvas 左上角的距离，Canvas 的左上角为原点，横向为 X 轴，纵向为 Y 轴	

3.6.9　target 与 currentTarget

构建项目的“最小程序状态”，在 index.js 文件中定义变量，参考代码如下，开发页面如图 3-60 所示。

```
fun0(event){
  console.log("fun0");
  console.log(JSON.stringify(event));
},
fun1(event){
  console.log("fun1");
  console.log(JSON.stringify(event));
},
```

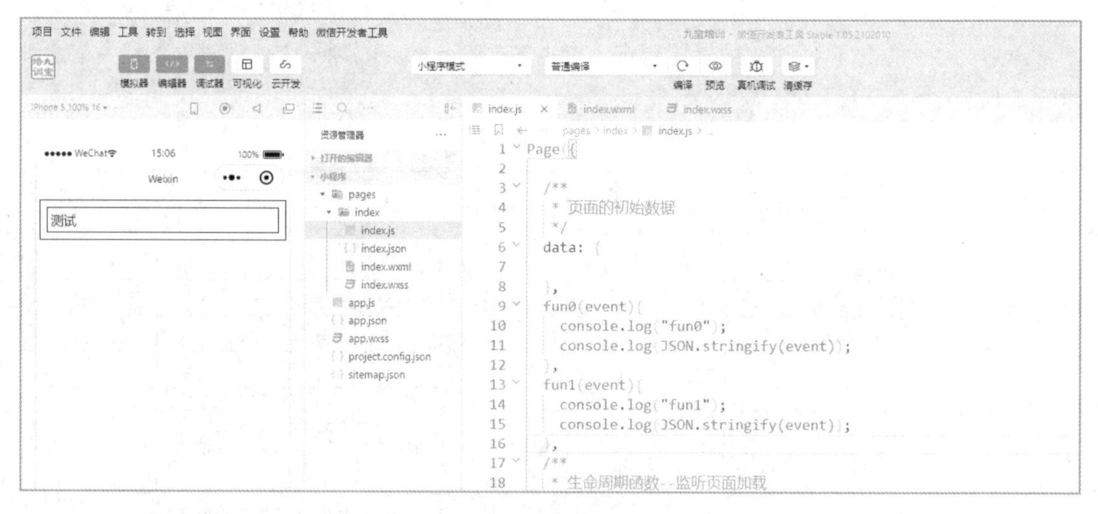

图 3-60

在 index.wxml 文件中绑定变量，参考代码如下，开发页面如图 3-61 所示。

```
<view bindtap="fun0">
  <view bindtap="fun1" data-arg="中国">测试</view>
</view>
```

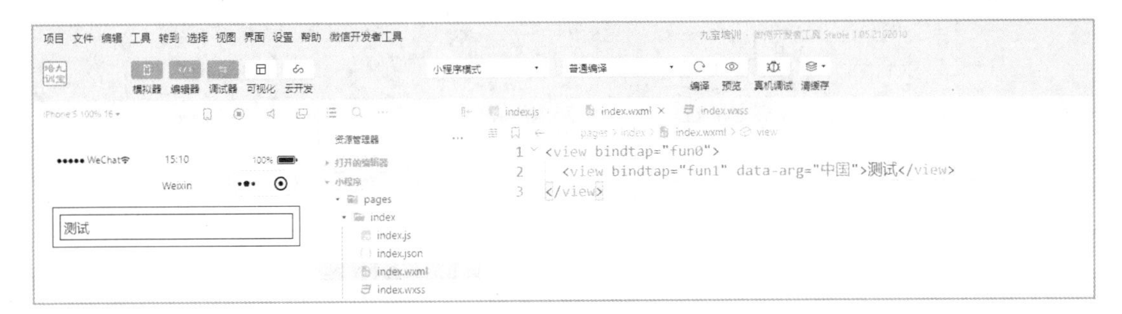

图 3-61

在 index.wxss 文件中绑定变量，参考代码如下，开发页面如图 3-62 所示。

```
view{
  margin: 10rpx;
  padding: 10rpx;
  border: 1px solid black;
}
```

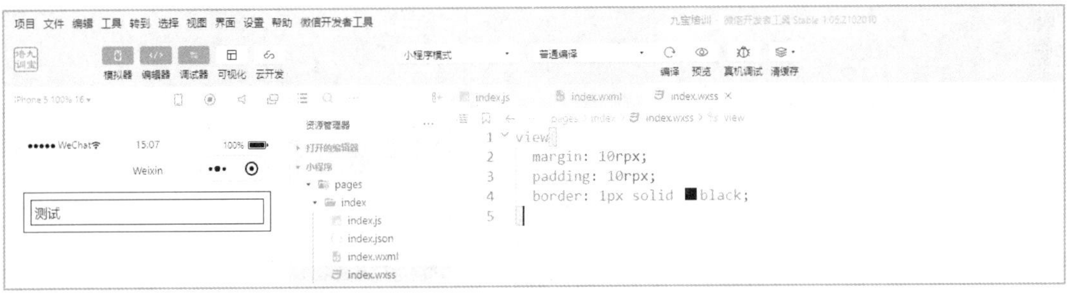

图 3-62

执行编译后，单击“测试”按钮，调试器打印信息如图 3-63 所示。

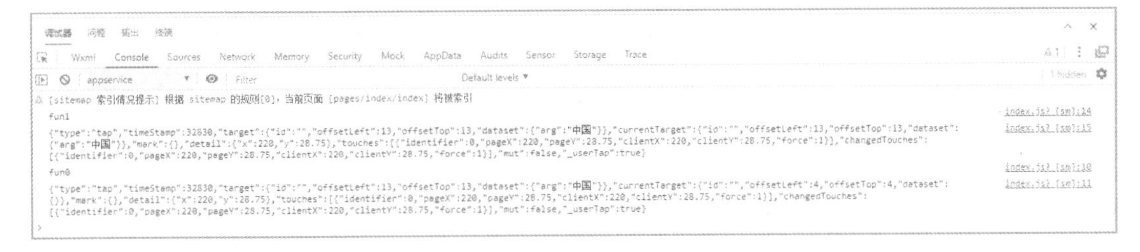

图 3-63

（1）fun1 的示例代码如下。

```
{
"type": "tap",
"timeStamp": 32830,
"target": {
    "id": "",
    "offsetLeft": 13,
    "offsetTop": 13,
    "dataset": {
        "arg": "中国"
    }
```

```
},
"currentTarget": {
    "id": "",
    "offsetLeft": 13,
    "offsetTop": 13,
    "dataset": {
        "arg": "中国"
    }
},
"mark": {},
"detail": {
    "x": 220,
    "y": 28.75
},
"touches": [{
    "identifier": 0,
    "pageX": 220,
    "pageY": 28.75,
    "clientX": 220,
    "clientY": 28.75,
    "force": 1
}],
"changedTouches": [{
    "identifier": 0,
    "pageX": 220,
    "pageY": 28.75,
    "clientX": 220,
    "clientY": 28.75,
    "force": 1
}],
"mut": false,
"_userTap": true
}
```

（2）fun0 的示例代码如下。

```
{
"type": "tap",
"timeStamp": 32830,
"target": {
    "id": "",
    "offsetLeft": 13,
    "offsetTop": 13,
    "dataset": {
        "arg": "中国"
    }
},
"currentTarget": {
    "id": "",
    "offsetLeft": 4,
    "offsetTop": 4,
    "dataset": {}
},
"mark": {},
"detail": {
    "x": 220,
    "y": 28.75
```

```
    },
    "touches": [{
        "identifier": 0,
        "pageX": 220,
        "pageY": 28.75,
        "clientX": 220,
        "clientY": 28.75,
        "force": 1
    }],
    "changedTouches": [{
        "identifier": 0,
        "pageX": 220,
        "pageY": 28.75,
        "clientX": 220,
        "clientY": 28.75,
        "force": 1
    }],
    "mut": false,
    "_userTap": true
    }
```

通过对比可以发现，按钮<view bindtap="fun1" data-arg="中国">测试</view>的 target 与 currentTarget 的打印数据是相同的。但对于外层<view>，target 有数据，currentTarget 无数据。target 是触发事件的组件的一些属性值集合，currentTarget 是当前组件的一些属性值集合。

第 4 章　小程序组件

4.1　概述

本章开始系统讲解微信小程序的各类组件。微信小程序组件共分为 9 类，具体类型及组件功能说明如表 4-1 所示。

表 4-1

类　型	名　称	功 能 说 明
视图容器组件	cover-image	覆盖在原生组件之上的图片视图
	cover-view	覆盖在原生组件之上的文本视图
	match-media	media query 匹配检测节点
	movable-area	movable-view 的可移动区域
	movable-view	可移动的视图容器，在页面中可以拖曳滑动
	page-container	页面容器
	scroll-view	可滚动视图区域
	share-element	共享元素
	swiper	滑块视图容器
	swiper-item	仅可放置在 swiper 组件中，宽和高自动设置为 100%
	view	视图容器
基础内容组件	icon	图标
	progress	进度条
	rich-text	富文本
	text	文本
表单组件	button	按钮
	checkbox	多选项目
	checkbox-group	多项选择器，内部由多个 checkbox 组成
	editor	富文本编辑器，可以对图片、文字进行编辑
	form	表单
	input	输入框
	keyboard-accessory	设置 input/textarea 聚焦时键盘上方 cover-view/cover-image 工具栏视图
	label	用来改进表单组件的可用性
	picker	从底部弹起的滚动选择器
	picker-view	嵌入页面的滚动选择器
	picker-view-column	滚动选择器子项
	radio	单选项目
	radio-group	单项选择器，内部由多个 radio 组成
	slider	滑动选择器
	switch	开关选择器
	textarea	多行输入框

续表

类　型	名　称	功 能 说 明
导航组件	functional-page-navigator	仅在插件中有效，用于跳转到插件功能页
	navigator	页面链接
媒体组件	audio	音频
	camera	系统相机
	image	图片
	live-player	实时音视频播放（v2.9.1 起支持同层渲染）
	live-pusher	实时音视频录制（v2.9.1 起支持同层渲染）
	video	视频（v2.4.0 起支持同层渲染）
	voip-room	多人音视频对话
地图组件	map	地图（从 v2.7.0 起支持同层渲染），相关 API：wx.createMapContext
画布组件	canvas	画布
开放能力组件	web-view	承载网页的容器
	ad	Banner（横幅）广告
	ad-custom	原生模板广告
	official-account	公众号关注组件
	open-data	用于展示微信开放的数据
原生组件	native-component	由客户端创建，包括 camera、canvas、video 等组件

本着简单、高效的原则，本节只选取比较典型的组件进行讲解。

4.2　视图容器组件

视图容器组件的典型组件主要包括 scroll-view 组件、share-element 与 page-container 组件、swiper 与 swiper-item 组件、view 组件，下面分别进行介绍。

4.2.1　scroll-view 组件

scroll-view 组件用于创建可滚动视图区域，实现效果如图 4-1 所示。

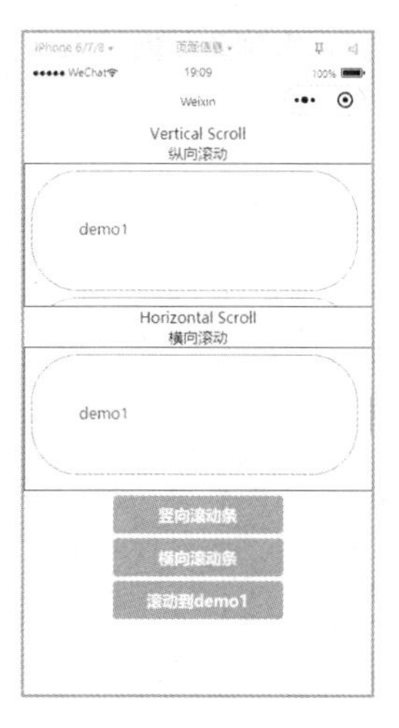

图 4-1

（1）JS 程序代码如下。

```
Page({

  data: {
    toView: '',
    scrollTop: '',
    scrollLeft: '',
  },

  scrolltoupper(event) {
    console.log("scrolltoupper" + JSON.stringify(event));
  },

  scrolltolower(event) {
    console.log("scrolltolower" + JSON.stringify(event));
  },

  scroll(event) {
    //console.log("scroll:" + JSON.stringify(event));
  },

  click_scrollTop() {
    this.setData({
      scrollTop: this.data.scrollTop * 1 + 10
    })
  },

  click_scrollLeft() {
    this.setData({
      scrollLeft: this.data.scrollLeft * 1 + 10
    })
  },

  click_toView() {
    this.setData({
      toView: 'demo1'
    });
  },

})
```

（2）WXML 程序代码如下。

```
<view class="page-section-title">
  <text>Vertical Scroll\n纵向滚动</text>
</view>
<view>
  <scroll-view scroll-y="true" style="height: 300rpx;"
  upper-threshold="10rpx"
  lower-threshold="10rpx"
  bindscrolltoupper="scrolltoupper"
  bindscrolltolower="scrolltolower"
  bindscroll="scroll"
  scroll-into-view="{{toView}}"
  scroll-top="{{scrollTop}}">
```

```
      <view id='demo1' class="demo" style="display: flex;"><view class="text">
demo1</view></view>
      <view id='demo2' class="demo" style="display: flex;"><view class="text">
demo2</view></view>
      <view id='demo3' class="demo" style="display: flex;"><view class="text">
demo3</view></view>
    </scroll-view>
  </view>

  <view class="page-section-title">
    <text>Horizontal Scroll\n 横向滚动</text>
  </view>
  <view>
    <scroll-view class="scroll-view_H" scroll-x="true" style="height: 300rpx;"
    upper-threshold="10rpx"
    lower-threshold="10rpx"
    bindscrolltoupper="scrolltoupper"
    bindscrolltolower="scrolltolower"
    bindscroll="scroll"
    scroll-into-view="{{toView}}"
    scroll-left="{{scrollLeft}}">
      <view id="demo1" class="demo"><view class="text">demo1</view></view>
      <view id="demo2" class="demo"><view class="text">demo2</view></view>
      <view id="demo3" class="demo"><view class="text">demo3</view></view>
    </scroll-view>
  </view>

  <view>
    <button type="primary" bindtap="click_scrollTop">竖向滚动条</button>
    <button type="primary" bindtap="click_scrollLeft">横向滚动条</button>
    <button type="primary" bindtap="click_toView">滚动到 demo1</button>

  </view>
```

（3）WXSS 程序代码如下。

```
  .scroll-view_H{
    white-space: nowrap;
  }

  .demo{
    display: inline-block;
    width: 700rpx;
    height: 250rpx;
    margin: 15rpx;
    border-radius:  100rpx 100rpx;
    border: 1px black dotted;
  }
  .text{
    margin-top: 100rpx;
    margin-left: 100rpx;
  }

  scroll-view{
    border: 1px black solid;
  }

  .page-section-title{text-align: center;}
  button{margin: 10rpx;}
```

scroll-view 组件可以实现屏幕的横向、纵向滚动，效果如图 4-2 所示。

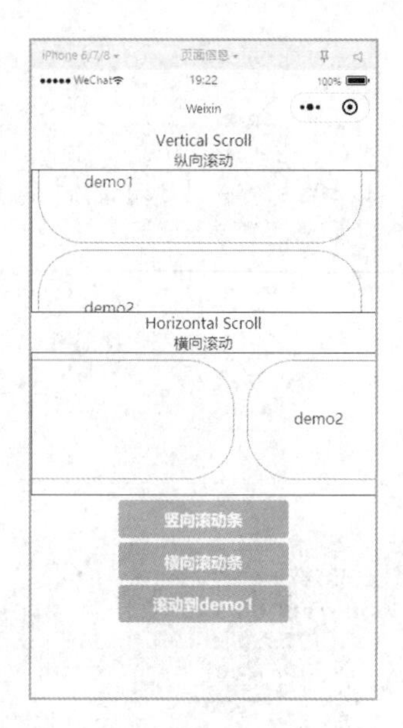

图 4-2

当屏幕滚动到距顶部/左边 upper-threshold 时，会触发 scrolltoupper 事件；滚动到底部/右边 lower-threshold 时，会触发 scrolltolower 事件。

- ◆ bindscroll：滚动时触发事件。
- ◆ scroll-into-view：定义滚动到指定 id 的组件。
- ◆ scroll-top/scroll-left：竖向/横向滚动条的位置。

相关示例及效果如图 4-3 所示。

图 4-3

4.2.2 share-element 与 page-container 组件

share-element 是共享元素，可实现类似于 Flutter Hero 动画在页面间穿越的效果。该组件需要与 page-container 组件结合起来使用。

使用该组件时，需在当前页面放置 share-element 组件，并在 page-container 容器中放置对应的 share-element 组件，对应关系通过属性值 key 进行映射。当设置 page-container 显示时，transform 属性为 true 的共享元素会产生动画。当前页面容器退出时，会产生返回动画。

share-element 组件的常见属性如表 4-2 所示。

表 4-2

属　性	类　型	默 认 值	必　填	说　明	最 低 版 本
key	string		是	映射标记	2.16.0
transform	boolean	false	否	是否进行动画	2.16.0
duration	number	300	否	动画时长，单位为毫秒	2.16.0
easing-function	string	ease-out	否	CSS 缓动函数	2.16.0

page-container 是页面容器，提供“假页”容器组件，效果类似于 popup 弹出层。当页面内存在该容器时，用户进行返回操作会关闭该容器，但不会关闭页面。返回操作包括右滑手势、Android 返回键和调用 navigateBack 接口 3 种情形。

page-container 组件的属性如表 4-3 所示。

表 4-3

属　性	类　型	默 认 值	必　填	说　明	最 低 版 本
show	boolean	false	否	是否显示容器组件	2.16.0
duration	number	300	否	动画时长，单位为毫秒	2.16.0
z-index	number	100	否	z-index 层级	2.16.0
overlay	boolean	true	否	是否显示遮罩层	2.16.0
position	string	bottom	否	弹出位置，可选值为 top、bottom、right 和 center	2.16.0
round	boolean	false	否	是否显示圆角	2.16.0
close-on-slideDown	boolean	false	否	是否在下滑一段距离后关闭	2.16.0
overlay-style	string		否	自定义遮罩层样式	2.16.0
custom-style	string		否	自定义弹出层样式	2.16.0
bind:beforeenter	eventhandle		否	进入前触发	2.16.0
bind:enter	eventhandle		否	进入中触发	2.16.0
bind:afterenter	eventhandle		否	进入后触发	2.16.0
bind:beforeleave	eventhandle		否	离开前触发	2.16.0
bind:leave	eventhandle		否	离开中触发	2.16.0
bind:afterleave	eventhandle		否	离开后触发	2.16.0
bind:clickoverlay	eventhandle		否	单击遮罩层时触发	2.16.0

下面将介绍 share-element 与 page-container 的示例程序，了解一下其功能。

（1）JS 程序代码如下。

```
<share-element key="share-element0" transform="{{false}}">
  share-element0
</share-element>
<share-element key="share-element1" transform="{{false}}">
  share-element1
</share-element>
<button type="primary"
 bindtap="show">share-element</button>
```

```
<page-container
  overlay='{{false}}'
  position='center'
  show="{{share_element_show}}"
  close-on-slideDown='{true}'>
  <view class="page_container">
    <share-element key="share-element0">
      page-container-share-element0
    </share-element>
    <share-element key="share-element1">
      page-container-share-element1
    </share-element>
  </view>
  <button type="primary"
 bindtap="show">share-element</button>
</page-container>
```

（2）WXML 程序代码如下。

```
Page({
  data: {
    share_element_show: false,
  },
  show() {
      this.setData({
        "share_element_show": !this.data.share_element_show
      });
  },
})
```

（3）WXSS 程序代码如下。

```
share-element{
  margin: 100rpx;
}

.page_container{
  border:10px black dotted;
}
```

编译程序，效果如图 4-4 所示。单击 share-element 按钮，效果如图 4-5 所示。

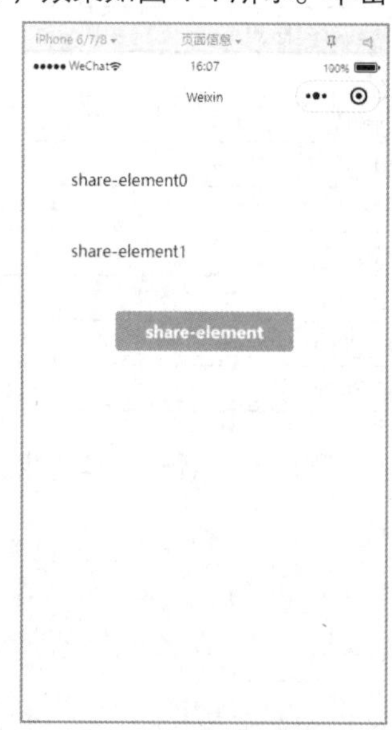

图 4-4

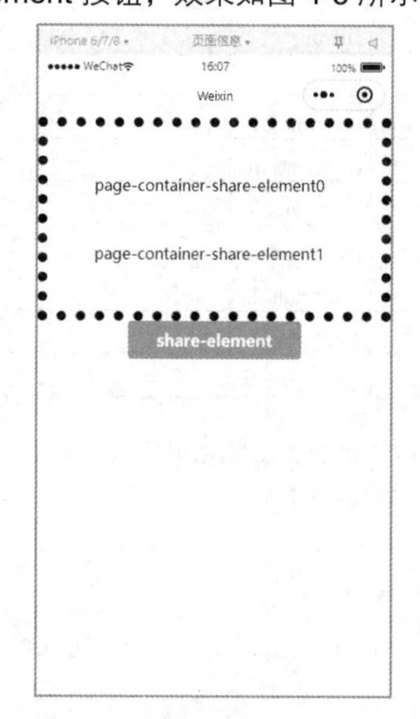

图 4-5

4.2.3 swiper 与 swiper-item 组件

swiper 与 swiper-item 组件的实现效果如图 4-6 所示。

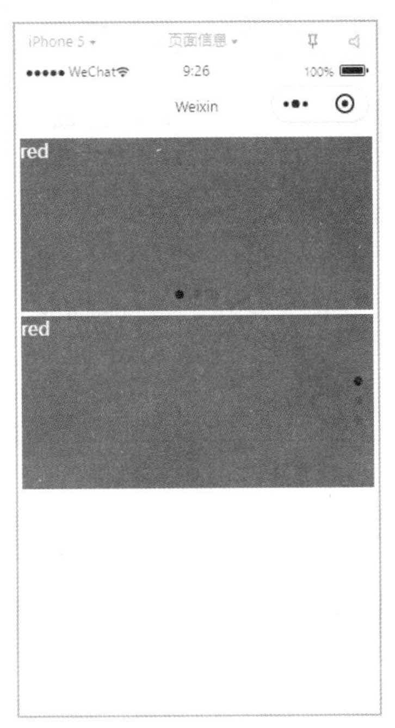

图 4-6

swiper 是滑块视图容器，其中只可放置 swiper-item 组件。swiper 组件的属性如表 4-4 所示。

表 4-4

属　　性	类　　型	默 认 值	必　　填	说　　明	最 低 版 本
indicator-dots	boolean	false	否	是否显示面板指示点	1.0.0
indicator-color	color	rgba(0,0,0,.3)	否	指示点颜色	1.1.0
indicator-active-color	color	#000000	否	当前选中的指示点颜色	1.1.0
autoplay	boolean	false	否	是否自动切换	1.0.0
current	number	0	否	当前所在滑块的 index	1.0.0
interval	number	5000	否	自动切换时间间隔	1.0.0
duration	number	500	否	滑动动画时长	1.0.0
circular	boolean	false	否	是否采用衔接滑动	1.0.0
vertical	boolean	false	否	滑动方向是否为纵向	1.0.0
previous-margin	string	"0px"	否	前边距，可用于露出前一项的一小部分，接受 px 和 rpx 值	1.9.0
next-margin	string	"0px"	否	后边距，可用于露出后一项的一小部分，接受 px 和 rpx 值	1.9.0
snap-to-edge	boolean	"false"	否	当 swiper-item 个数大于等于 2，关闭 circular 且开启 previous-margin 或 next-margin 时，可指定该边距是否应用第一个、最后一个元素	2.12.1

续表

属　性	类　型	默 认 值	必　填	说　明	最 低 版 本
display-multiple-items	number	1	否	同时显示的滑块数量	1.9.0
easing-function	string	"default"	否	指定 swiper 切换缓动动画类型。 default：默认缓动函数； linear：线性动画； easeInCubic：缓入动画； easeOutCubic：缓出动画； easeInOutCubic：缓入缓出动画	2.6.5
bindchange	eventhandle		否	current 改变时会触发 change 事件，event.detail = {current, source}	1.0.0
bindtransition	eventhandle		否	swiper-item 的位置发生改变时会触发 transition 事件，event.detail = {dx: dx, dy: dy}	2.4.3
bindanimationfinish	eventhandle		否	动画结束时会触发 animationfinish 事件，event.detail 同上	1.9.0

（1）swiper 与 swiper-item 组件没有 JS 程序。

（2）WXML 程序示例代码如下。

```
<swiper indicator-dots autoplay interval='1000' circular>
  <swiper-item>
    <view class="red">red</view>
  </swiper-item>
  <swiper-item>
    <view class="green">green</view>
  </swiper-item>
  <swiper-item>
    <view class="blue">blue</view>
  </swiper-item>
</swiper>

<swiper indicator-dots autoplay interval='1000' vertical>
  <swiper-item>
    <view class="red">red</view>
  </swiper-item>
  <swiper-item>
    <view class="green">green</view>
  </swiper-item>
  <swiper-item>
    <view class="blue">blue</view>
  </swiper-item>
</swiper>
```

（3）WXSS 程序代码如下。

```
.red{
  background: red;
}

.green{
  background: green;
}
```

```
.blue{
  background: blue;
}

view{
  height: 100%;
  width: 100%;
  color: #fff;
}

swiper{
  margin: 10rpx;
}
```

编译小程序后的效果如图 4-7~图 4-9 所示。

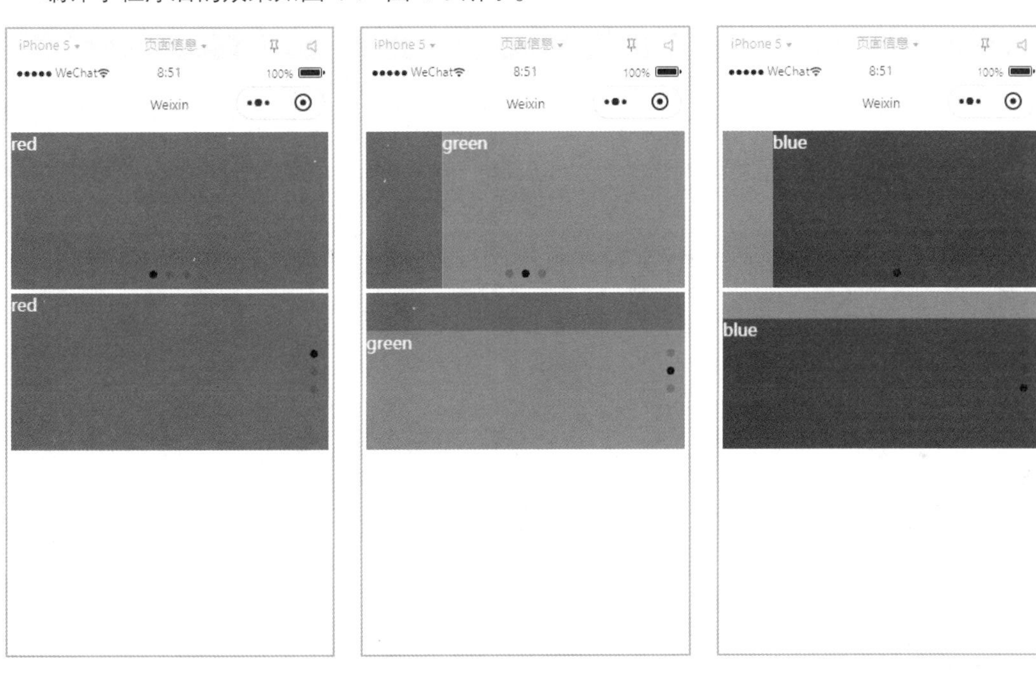

图 4-7　　图 4-8　　图 4-9

4.2.4 view 组件

view 是视图容器，相当于 HTML 中的<div>标签，其属性如表 4-5 所示。

表 4-5

属　性	类　型	默 认 值	必　填	说　明	最 低 版 本
hover-class	string	none	否	指定按下去的样式类。当 hover-class="none" 时，没有单击态效果	1.0.0
hover-stop-propagation	boolean	false	否	指定是否阻止本节点的祖先节点出现单击态	1.5.0
hover-start-time	number	50	否	按住后多久出现单击态，单位为毫秒	1.0.0
hover-stay-time	number	400	否	手指松开后单击态保留时间，单位为毫秒	1.0.0

（1）view 组件没有 JS 程序。

（2）WXML 程序示例代码如下。

```
<view class='_view'>
  <view class='demo'>1</view>
  <view class='demo'>2</view>
  <view class='demo'>3</view>
</view>

<view class='demo' hover-class='click_class'>
  Test
</view>
```

（3）WXSS 程序示例代码如下。

```
._view{
  display: flex;
}

.demo{
  flex: 1;
  border: 1px solid black;
  font-size: xx-large;
}

.click_class{
  background: blue;
  color: white;
}
```

编译小程序，效果如图 4-10 所示。单击 Test 按钮，效果如图 4-11 所示。恢复原始效果，如图 4-12 所示。

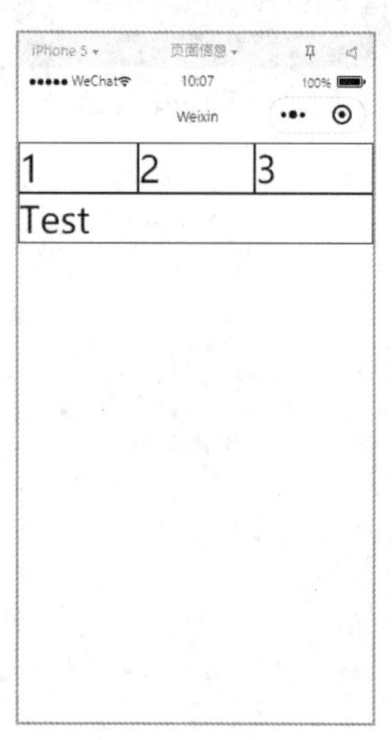

图 4-10

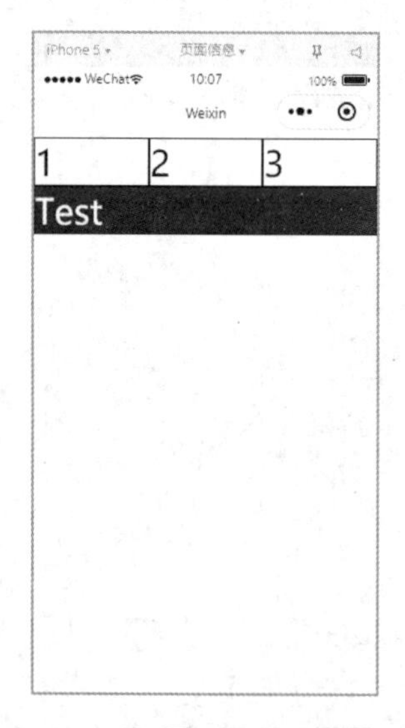

图 4-11

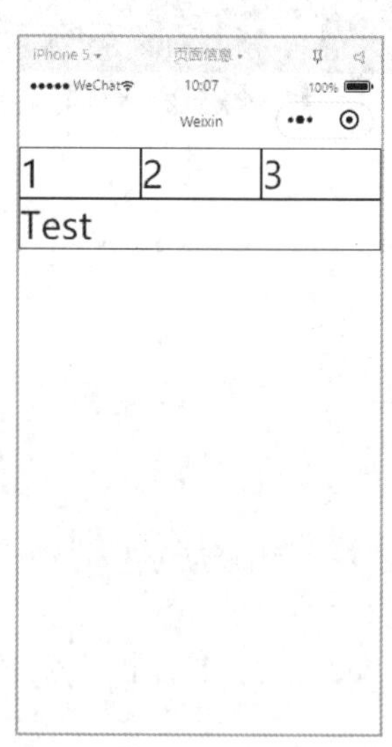

图 4-12

4.3 基础内容组件

基础内容组件的典型组件主要包括 icon 组件、progress 组件、rich-text 组件、text 组件，下面分别进行介绍。

4.3.1　icon 组件

icon 是图标组件，其效果如图 4-13 所示。

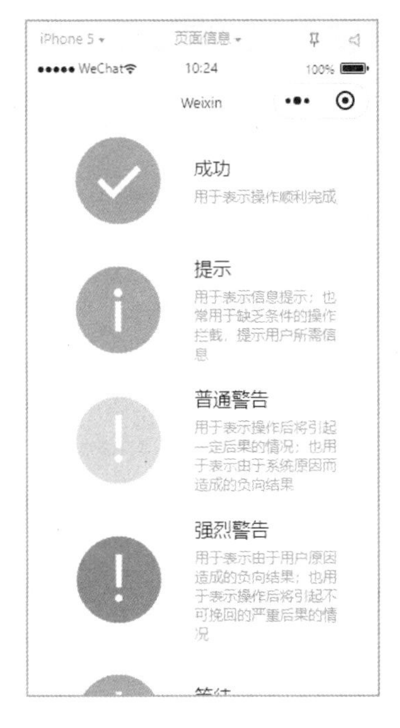

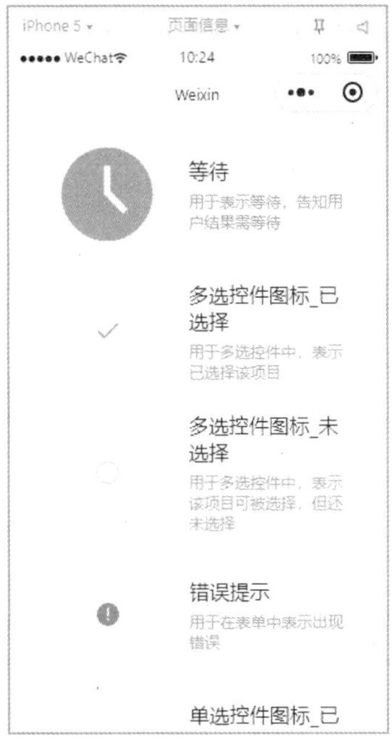

图 4-13

（1）icon 组件没有 JS 程序。

（2）WXML 示例代码如下。

```
    <view class="icon-box">
      <icon class="icon-box-img" type="success" size="93"></icon>
      <view class="icon-box-ctn">
        <view class="icon-box-title">成功</view>
        <view class="icon-box-desc">用于表示操作顺利完成</view>
      </view>
    </view>
    <view class="icon-box">
      <icon class="icon-box-img" type="info" size="93"></icon>
      <view class="icon-box-ctn">
        <view class="icon-box-title">提示</view>
        <view class="icon-box-desc">用于表示信息提示；也常用于缺乏条件的操作拦截，提示用户所需信
息</view>
      </view>
    </view>
    <view class="icon-box">
      <icon class="icon-box-img" type="warn" size="93" color="#C9C9C9"></icon>
      <view class="icon-box-ctn">
        <view class="icon-box-title">普通警告</view>
        <view class="icon-box-desc">用于表示操作后将引起一定后果的情况；也用于表示由于系统原因而
造成的负向结果</view>
      </view>
    </view>
    <view class="icon-box">
      <icon class="icon-box-img" type="warn" size="93"></icon>
```

```
    <view class="icon-box-ctn">
      <view class="icon-box-title">强烈警告</view>
      <view class="icon-box-desc">用于表示由于用户原因造成的负向结果；也用于表示操作后将引起不可挽回的严重后果的情况</view>
    </view>
  </view>
  <view class="icon-box">
    <icon class="icon-box-img" type="waiting" size="93"></icon>
    <view class="icon-box-ctn">
      <view class="icon-box-title">等待</view>
      <view class="icon-box-desc">用于表示等待，告知用户结果需等待</view>
    </view>
  </view>
  <view class="icon-box">
    <view class="icon-small-wrp">
      <icon class="icon-small" type="success_no_circle" size="23"></icon>
    </view>
    <view class="icon-box-ctn">
      <view class="icon-box-title">多选控件图标_已选择</view>
      <view class="icon-box-desc">用于多选控件中，表示已选择该项目</view>
    </view>
  </view>
  <view class="icon-box">
    <view class="icon-small-wrp">
      <icon class="icon-small" type="circle" size="23"></icon>
    </view>
    <view class="icon-box-ctn">
      <view class="icon-box-title">多选控件图标_未选择</view>
      <view class="icon-box-desc">用于多选控件中，表示该项目可被选择，但还未选择</view>
    </view>
  </view>
  <view class="icon-box">
    <view class="icon-small-wrp">
      <icon class="icon-small" type="warn" size="23"></icon>
    </view>
    <view class="icon-box-ctn">
      <view class="icon-box-title">错误提示</view>
      <view class="icon-box-desc">用于在表单中表示出现错误</view>
    </view>
  </view>
  <view class="icon-box">
    <view class="icon-small-wrp">
      <icon class="icon-small" type="success" size="23"></icon>
    </view>
    <view class="icon-box-ctn">
      <view class="icon-box-title">单选控件图标_已选择</view>
      <view class="icon-box-desc">用于单选控件中，表示已选择该项目</view>
    </view>
  </view>
  <view class="icon-box">
    <view class="icon-small-wrp">
      <icon class="icon-small" type="download" size="23"></icon>
    </view>
    <view class="icon-box-ctn">
      <view class="icon-box-title">下载</view>
      <view class="icon-box-desc">用于表示可下载</view>
    </view>
  </view>
  <view class="icon-box">
    <view class="icon-small-wrp">
      <icon class="icon-small" type="info_circle" size="23"></icon>
    </view>
```

```
    <view class="icon-box-ctn">
      <view class="icon-box-title">提示</view>
      <view class="icon-box-desc">用于在表单中表示有信息提示</view>
    </view>
  </view>
  <view class="icon-box">
    <view class="icon-small-wrp">
      <icon class="icon-small" type="cancel" size="23"></icon>
    </view>
    <view class="icon-box-ctn">
      <view class="icon-box-title">停止或关闭</view>
      <view class="icon-box-desc">用于在表单中表示关闭或停止</view>
    </view>
  </view>
  <view class="icon-box">
    <view class="icon-small-wrp">
      <icon class="icon-small" type="search" size="14"></icon>
    </view>
    <view class="icon-box-ctn">
      <view class="icon-box-title">搜索</view>
      <view class="icon-box-desc">用于搜索控件中表示可搜索</view>
    </view>
  </view>
```

（3）WXSS 示例代码如下。

```
    .icon-box{ margin-bottom: 20px; padding: 0 37px; display: flex; align-items: center; }
    .icon-box-img{ margin-right: 23px; }
    .icon-box-ctn{ flex-shrink: 100; }
    .icon-box-title{ font-size: 17px; }
    .icon-box-desc{ margin-top: 6px; font-size: 13px; color: #888; }
    .icon-small-wrp{ margin-right: 23px; width: 93px; height: 93px; display: flex;
align-items: center; justify-content: center; }
```

4.3.2　progress 组件

progress 是进度条组件，其效果如图 4-14 所示。

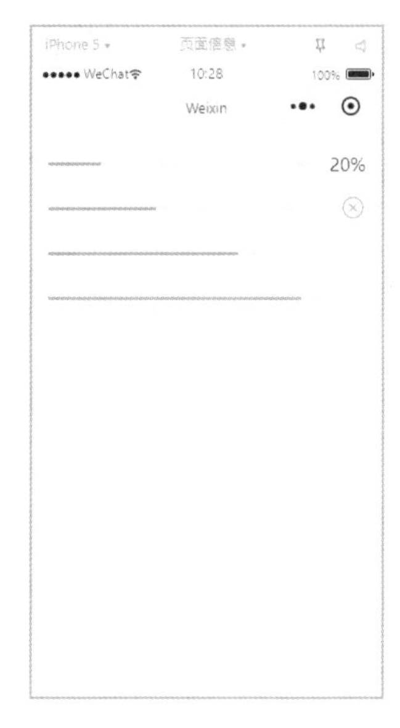

图 4-14

（1）progress 组件没有 JS 程序。

（2）WXML 程序示例代码如下。

```
<view class="progress-box">
  <progress percent="20" show-info stroke-width="3"/>
</view>

<view class="progress-box">
  <progress percent="40" active stroke-width="3" />
  <icon class="progress-cancel" type="cancel"></icon>
</view>

<view class="progress-box">
  <progress percent="60" active stroke-width="3" />
</view>

<view class="progress-box">
  <progress percent="80" color="#10AEFF" active stroke-width="3" />
</view>
```

（3）WXSS 程序示例代码如下。

```
wx-progress{ width: 100%; }
.progress-box { display: flex; height: 25px; margin: 15px; }
.progress-cancel{ margin-left: 20px; }
```

4.3.3 rich-text 组件

rich-text 是富文本组件，其效果如图 4-15 所示。

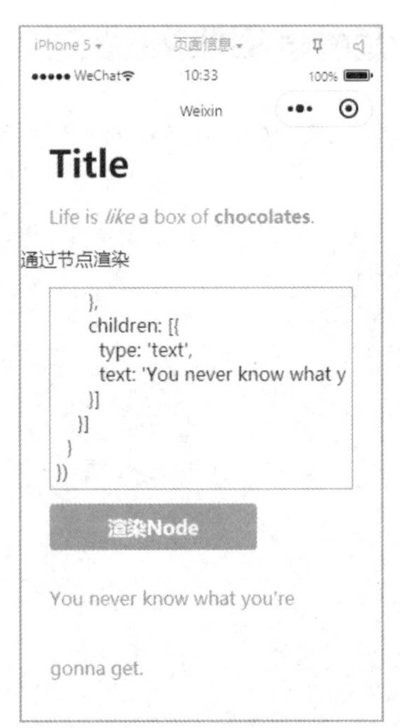

图 4-15

（1）JS 程序示例代码如下。

```
const htmlSnip =
`<div class="div_class">
  <h1>Title</h1>
  <p class="p">
    Life is <i>like</i> a box of
    <b> chocolates</b>.
  </p>
</div>
`

const nodeSnip =
`Page({
  data: {
    nodes: [{
      name: 'div',
      attrs: {
        class: 'div_class',
        style: 'line-height: 60px; color: red;'
      },
      children: [{
        type: 'text',
        text: 'You never know what you're gonna get.'
      }]
    }]
  }
})
`

Page({
  onShareAppMessage() {
    return {
      title: 'rich-text',
      path: 'page/component/pages/rich-text/rich-text'
    }
  },

  data: {
    htmlSnip,
    nodeSnip,
    renderedByHtml: false,
    renderedByNode: false,
    nodes: [{
      name: 'div',
      attrs: {
        class: 'div_class',
        style: 'line-height: 60px; color: #1AAD19;'
      },
      children: [{
        type: 'text',
        text: 'You never know what you\'re gonna get.'
      }]
    }]
  },
  renderHtml() {
```

```
      this.setData({
        renderedByHtml: true
      })
    },
    renderNode() {
      this.setData({
        renderedByNode: true
      })
    },
    enterCode(e) {
      console.log(e.detail.value)
      this.setData({
        htmlSnip: e.detail.value
      })
    }
  })
```

（2）WXML 程序示例代码如下。

```
  <view>
    <view class="page-body">
      <view class="page-section">
        <view class="page-section-title">通过 HTML String 渲染</view>
        <view class="page-content">
          <scroll-view scroll-y>{{htmlSnip}}</scroll-view>
          <button style="margin: 30rpx 0" type="primary" bindtap="renderHtml">渲染
HTML</button>
          <block wx:if="{{renderedByHtml}}">
            <rich-text nodes="{{htmlSnip}}"></rich-text>
          </block>
        </view>
      </view>

      <view class="page-section">
        <view class="page-section-title">通过节点渲染</view>
        <view class="page-content">
          <scroll-view scroll-y>{{nodeSnip}}</scroll-view>
          <button style="margin: 30rpx 0" type="primary" bindtap="renderNode">渲染
Node</button>
          <block wx:if="{{renderedByNode}}">
            <rich-text nodes="{{nodes}}"></rich-text>
          </block>
        </view>
      </view>
    </view>

  </view>
```

（3）WXSS 程序示例代码如下。

```
  .page-content { width: auto; margin: 15px 0; padding: 0 25px; }
  .code-fragment { height: 175px; }
  .code-textarea { width: auto; height: 175px; border: 1px solid #1AAD19; }
  .p { color: #1AAD19; margin-top: 15px; }
  .space { display: inline-block; background: red; width: 5px; }
  wx-scroll-view { height: 175px; border: 1px solid #1AAD19; white-space: pre; padding:
5px; box-sizing: border-box; }
```

编译后效果如图 4-16 所示。单击“渲染 HTML”按钮，效果如图 4-17 所示。单击“渲染 Node”按钮，效果如图 4-18 所示。

图 4-16

图 4-17

图 4-18

4.3.4　text 组件

（1）JS 程序示例代码如下。

```
const texts = [
  '桃花源记',
  '陶渊明〔魏晋〕',
  '晋太元中，武陵人捕鱼为业。',
  '缘溪行，忘路之远近。',
  '忽逢桃花林，夹岸数百步，中无杂树，芳草鲜美，落英缤纷。',
  '渔人甚异之，复前行，欲穷其林。',
  '林尽水源，便得一山，山有小口，仿佛若有光。',
  '便舍船，从口入。',
  '初极狭，才通人。',
  '复行数十步，豁然开朗。',
  '土地平旷，屋舍俨然，有良田、美池、桑竹之属。',
  '阡陌交通，鸡犬相闻。',
  '其中往来种作，男女衣着，悉如外人。',
  '黄发垂髫，并怡然自乐。',
  '见渔人，乃大惊，问所从来。',
  '具答之。',
  '便要还家，设酒杀鸡作食。',
  '村中闻有此人，咸来问讯。',
  '自云先世避秦时乱，率妻子邑人来此绝境，不复出焉，遂与外人间隔。',
  '问今是何世，乃不知有汉，无论魏晋。',
  '此人一一为具言所闻，皆叹惋。',
  '余人各复延至其家，皆出酒食。',
  '停数日，辞去。',
  '此中人语云："不足为外人道也。"',
  '既出，得其船，便扶向路，处处志之。',
  '及郡下，诣太守，说如此。',
  '太守即遣人随其往，寻向所志，遂迷，不复得路。',
  '南阳刘子骥，高尚士也，闻之，欣然规往。',
```

```
  '未果，寻病终。',
  '后遂无问津者。'
]

Page({
  onShareAppMessage() {
    return {
      title: 'text',
      path: 'page/component/pages/text/text'
    }
  },

  data: {
    text: '',
    canAdd: true,
    canRemove: false
  },
  extraLine: [],

  add() {
    this.extraLine.push(texts[this.extraLine.length % texts.length])
    this.setData({
      text: this.extraLine.join('\n'),
      canAdd: this.extraLine.length < texts.length,
      canRemove: this.extraLine.length > 0
    })
    setTimeout(() => {
      this.setData({
        scrollTop: 99999
      })
    }, 0)
  },
  remove() {
    if (this.extraLine.length > 0) {
      this.extraLine.pop()
      this.setData({
        text: this.extraLine.join('\n'),
        canAdd: this.extraLine.length < texts.length,
        canRemove: this.extraLine.length > 0,
      })
    }
    setTimeout(() => {
      this.setData({
        scrollTop: 99999
      })
    }, 0)
  }
})
```

（2）WXML 程序示例代码如下。

```
<view class="container">
  <view class="page-body">
    <view class="page-section page-section-spacing">
      <view class="text-box" scroll-y="true" scroll-top="{{scrollTop}}">
        <text>{{text}}</text>
      </view>
      <button disabled="{{!canAdd}}" bindtap="add">add line</button>
      <button disabled="{{!canRemove}}" bindtap="remove">remove line</button>
    </view>
  </view>
</view>
```

（3）text 组件没有 WXSS 程序。

编译后的初始效果如图 4-19 所示。此时 add line 按钮可以使用，remove line 按钮禁止使用。单击 add line 按钮，陶渊明的《桃花源记》将逐行显示，如图 4-20 和图 4-21 所示。此时，remove line 按钮转为非禁用状态。

图 4-19

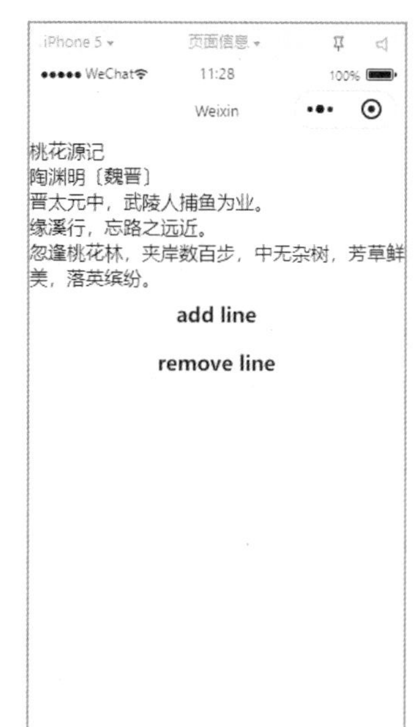

图 4-20

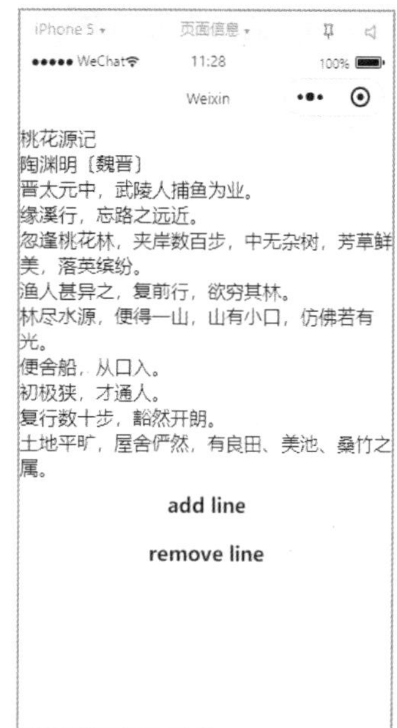

图 4-21

全文展示完毕后，add line 按钮将被禁用，如图 4-22 所示。

单击 remove line 按钮，将逐行删除古文内容，如图 4-23 所示。

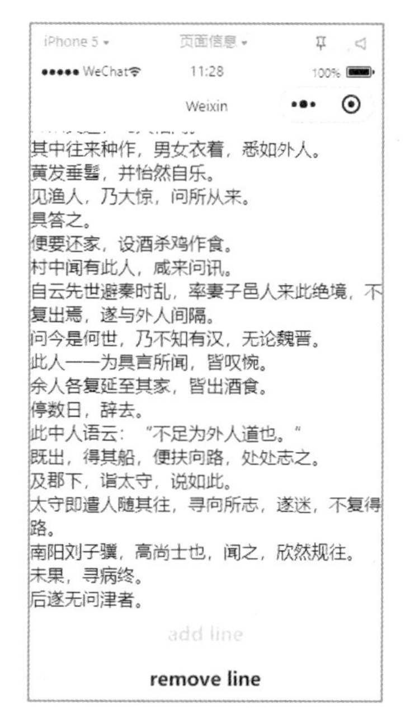

图 4-22

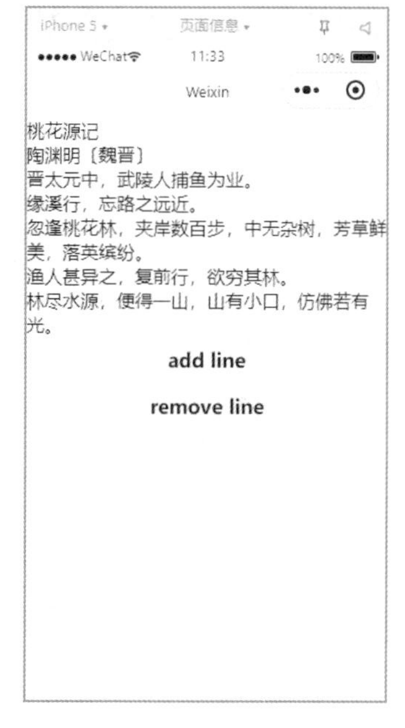

图 4-23

4.4 表单组件

与 HTML 不同，微信小程序中的表单组件没有“提交”功能。要想实现表单提交，需要在 button 中使用 form-type="submit" 代码进行设置。例如：

```
<button type="primary" size="mini" form-type="submit">submit</button>
```

此外，若想实现微信小程序中表单的重置功能，需要在 button 中使用 form-type="reset" 代码进行设置。例如：

```
<button type="primary" size="mini" form-type="reset">reset</button>
```

下面就结合微信小程序表单组件进行系统讲解。

（1）JS 程序代码如下。

```
Page({

  /**
   * 页面的初始数据
   */
  data: {

  },

  fun_bindsubmit(event){
    console.log('fun_bindsubmit');
    console.log(event);
  },

  fun_bindreset(event){
    console.log('fun_bindreset');
    console.log(event);
  },

  /**
   * 生命周期函数--监听页面加载
   */
  onLoad: function (options) {

  },

  /**
   * 生命周期函数--监听页面初次渲染完成
   */
  onReady: function () {

  },

  /**
   * 生命周期函数--监听页面显示
   */
  onShow: function () {
```

```
  },

  /**
   * 生命周期函数--监听页面隐藏
   */
  onHide: function () {

  },

  /**
   * 生命周期函数--监听页面卸载
   */
  onUnload: function () {

  },

  /**
   * 页面相关事件处理函数--监听用户下拉动作
   */
  onPullDownRefresh: function () {

  },

  /**
   * 页面上拉触底事件的处理函数
   */
  onReachBottom: function () {

  },

  /**
   * 用户单击右上角分享
   */
  onShareAppMessage: function () {

  }
})
```

（2）WXSS 程序代码如下。

```
view,button{
  margin: 50rpx;
}

input,textarea{
  border: 1px solid black;
  margin: 50rpx;
}

keyboard-accessory{
  display: flex;
}
```

4.4.1　form 组件

form 是最基本的表单组件。当单击 form 表单中 form-type 为 submit 的 button 组件时，会将表单组件中的 value 值进行提交。支持的组件有 switch、input、checkbox、slider、radio、picker 等。提交的组件需在表单组件中加上 name 来作为 key。form 组件属性如表 4-6 所示。

表 4-6

属　　性	类　　型	默 认 值	必　　填	说　　明	最 低 版 本
report-submit	boolean	false	否	是否返回 formId，用于发送模板消息	1.0.0
report-submit-timeout	number	0	否	等待一段时间（毫秒数），以确认 formId 是否生效。如果未指定该参数，formId 可能会无效（如网络失败）。指定该参数后可检测 formId 是否有效，以该参数的时间作为检测超时时间。失败则返回 requestFormId:fail 开头的 formId	2.6.2
bindsubmit	eventhandle		否	携带 form 中的数据触发 submit 事件，event.detail = {value : {'name': 'value'} , formId: ''}	1.0.0
bindreset	eventhandle		否	表单重置时会触发 reset 事件	1.0.0

WXML 程序代码如下。

```
<form bindsubmit='fun_bindsubmit' bindreset='fun_bindreset'>
  <view>最简单的 form
    <input name='userName' />
  </view>
  <button type="primary" size="mini" form-type="submit">submit</button>
  <button type="primary" size="mini" form-type="reset">reset</button>

</form>
```

编译后效果如图 4-24 所示。

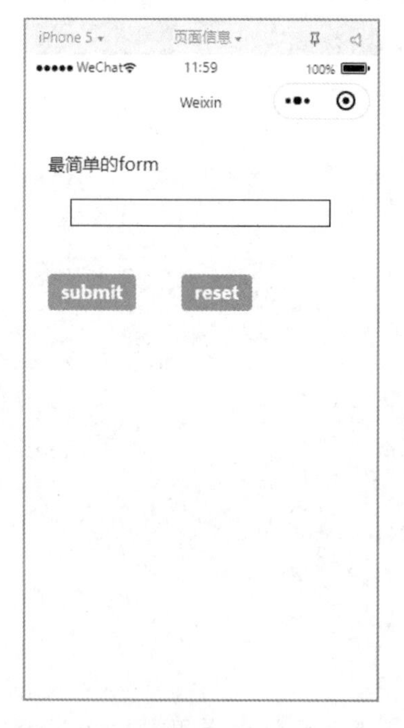

图 4-24

编译程序，在文本框中输入“九宝老师真帅”，单击 submit 按钮，Console 将打印表单数据，如图 4-25 所示。

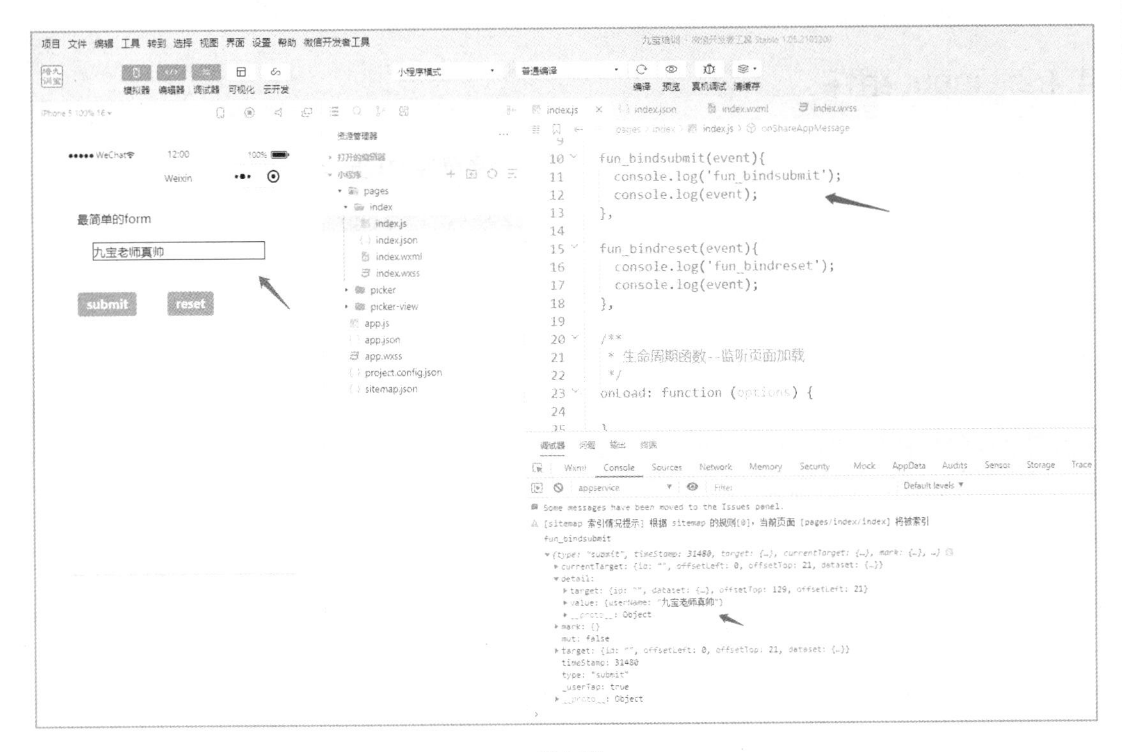

图 4-25

对于只关注表单数据的情况，可以修改 fun_bindsubmit 函数。

```
fun_bindsubmit(event){
  console.log('fun_bindsubmit');
  console.log(event.detail.value);
},
```

编译程序，在文本框中输入“九宝老师真帅”，单击 submit 按钮，Console 将打印表单数据，如图 4-26 所示。

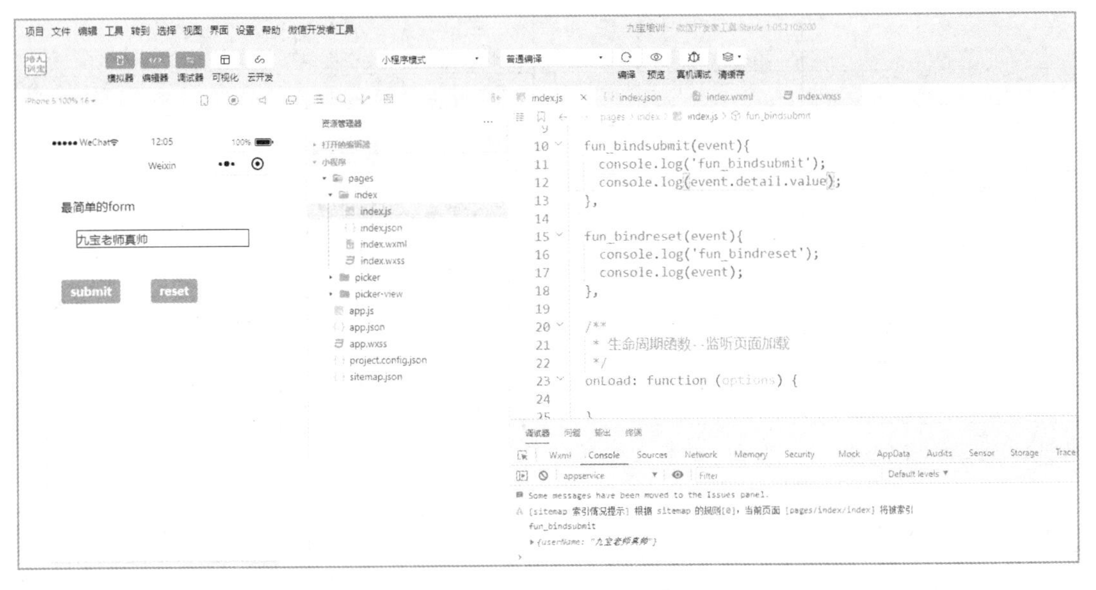

图 4-26

4.4.2 input 组件

input 是微信小程序表单组件中相对简单的组件，其属性如表 4-7 所示。

表 4-7

属性	类型	默认值	必填	说明	最低版本
value	string		是	输入框的初始内容	1.0.0
type	string	text	否	input 的类型。 text：文本输入； number：数字输入； idcard：身份证输入； digit：带小数点的数字输入	1.0.0
password	boolean	false	否	是否是密码类型	1.0.0
placeholder	string		是	输入框为空时的占位符	1.0.0
placeholder-style	string		是	指定 placeholder 的样式	1.0.0
placeholder-class	string	input-placeholder	否	指定 placeholder 的样式类	1.0.0
disabled	boolean	false	否	是否禁用	1.0.0
maxlength	number	140	否	最大输入长度，设置为-1 值时不限制最大长度	1.0.0
cursor-spacing	number	0	否	指定光标与键盘的距离，取 input 距离底部的距离和 cursor-spacing 指定的距离的最小值作为光标与键盘的距离	1.0.0
auto-focus	boolean	false	否	（即将废弃，请直接使用 focus）自动聚焦，拉起键盘	1.0.0
focus	boolean	false	否	获取焦点	1.0.0
confirm-type	string	done	否	设置键盘右下角按钮的文字，仅在 type='text'时生效。 send：按钮文字为“发送”； search：按钮文字为“搜索”； next：按钮文字为“下一个”； go：按钮文字为“前往”； done：按钮文字为“完成”	1.1.0
always-embed	boolean	false	否	强制 input 处于同层状态，默认 focus 时 input 会切换到非同层状态（仅在 iOS 下生效）	2.10.4
confirm-hold	boolean	false	否	单击键盘右下角按钮时是否保持键盘不收起	1.1.0
cursor	number		是	指定聚焦时的光标位置	1.5.0
selection-start	number	-1	否	光标起始位置，自动聚集时有效，需与 selection-end 搭配使用	1.9.0
selection-end	number	-1	否	光标结束位置，自动聚集时有效，需与 selection-start 搭配使用	1.9.0
adjust-position	boolean	true	否	键盘弹起时是否自动上推页面	1.9.90
hold-keyboard	boolean	false	否	聚焦时单击页面不收起键盘	2.8.2
bindinput	eventhandle		是	键盘输入时触发，event.detail = {value, cursor, keyCode}，keyCode 为键值，2.1.0 起支持，处理函数可以直接返回一个字符串，将替换输入框的内容	1.0.0

续表

属 性	类 型	默 认 值	必 填	说 明	最 低 版 本
bindfocus	eventhandle		是	输入框聚焦时触发，event.detail = { value, height }，height 为键盘高度，在基础库 1.9.90 起支持	1.0.0
bindblur	eventhandle		是	输入框失去焦点时触发，event.detail = {value: value}	1.0.0
bindconfirm	eventhandle		是	单击“完成”按钮时触发，event.detail = {value: value}	1.0.0
bindkeyboardheightchange	eventhandle		是	键盘高度发生变化时触发此事件，event.detail = {height: height, duration: duration}	2.7.0

WXML 程序的示例代码如下。

```
<form bindsubmit='fun_bindsubmit' bindreset='fun_bindreset'>

  <view>
    占位符
    <input placeholder='占位符' type='text' name='userName0' />
  </view>

  <view>
    input 的默认值
    <input type='text' name='userName1' value='默认值' />
  </view>

  <view>
    数字键盘
    <input type='number' name='userName2' value='000' />
  </view>

  <view>
    设置键盘右下角按钮的文字
    <input confirm-type='next' type='text' name='userName3' value='000' />
  </view>

  <button type="primary" size="mini" form-type="submit">submit</button>
  <button type="primary" size="mini" form-type="reset">reset</button>

</form>
```

在上述代码中，placeholder 用于定义输入框为空时的占位符；value 用于定义输入框的初始内容；type 用于定义类型；confirm-type 用于设置键盘右下角按钮的文字，仅在 type='text' 时生效。

程序编译后的效果如图 4-27 所示。数字键盘如图 4-28 所示。

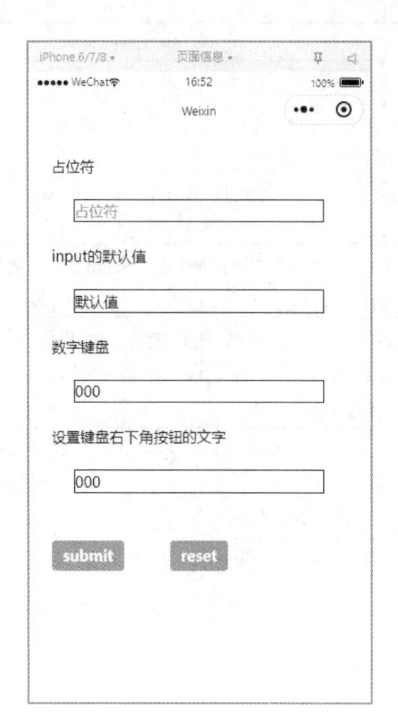

图 4-27

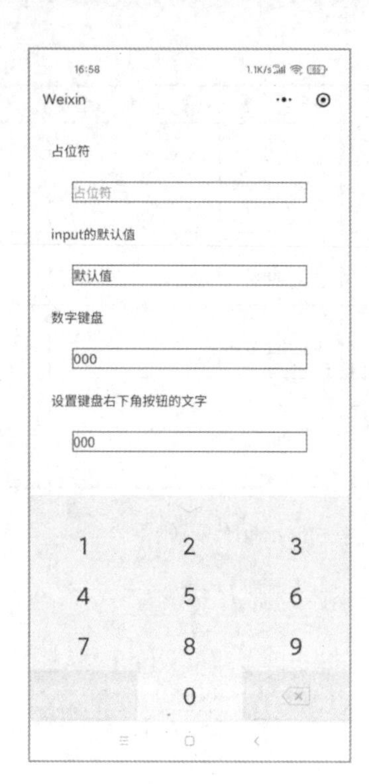

图 4-28

4.4.3 textarea 组件

textarea 是多行输入框组件，其属性如表 4-8 所示。

表 4-8

属　　性	类　　型	默 认 值	必　　填	说　　明	最 低 版 本
value	string		否	输入框的内容	1.0.0
placeholder	string		否	输入框为空时的占位符	1.0.0
placeholder-style	string		否	指定 placeholder 的样式，目前仅支持 color、font-size 和 font-weight	1.0.0
placeholder-class	string	textarea-placeholder	否	指定 placeholder 的样式类	1.0.0
disabled	boolean	false	否	是否禁用	1.0.0
maxlength	number	140	否	最大输入长度，设置为-1 值时不限制最大长度	1.0.0
auto-focus	boolean	false	否	自动聚焦，拉起键盘	1.0.0
focus	boolean	false	否	获取焦点	1.0.0
auto-height	boolean	false	否	是否自动增高，设置 auto-height 时，style.height 不生效	1.0.0
fixed	boolean	false	否	如果 textarea 在 position:fixed 区域，需要显示指定属性 fixed 为 true	1.0.0
cursor-spacing	number	0	否	指定光标与键盘的距离。取 textarea 距离底部的距离和 cursor-spacing 指定的距离的最小值作为光标与键盘的距离	1.0.0

续表

属　　性	类　　型	默 认 值	必　　填	说　　明	最 低 版 本
cursor	number	-1	否	指定 focus 时的光标位置	1.5.0
show-confirm-bar	boolean	true	否	是否显示键盘上方带有“完成”按钮那一栏	1.6.0
selection-start	number	-1	否	光标起始位置，自动聚集时有效，需与 selection-end 搭配使用	1.9.0
selection-end	number	-1	否	光标结束位置，自动聚集时有效，需与 selection-start 搭配使用	1.9.0
adjust-position	boolean	true	否	键盘弹起时是否自动上推页面	1.9.90
hold-keyboard	boolean	false	否	focus 时，单击页面时不收起键盘	2.8.2
disable-default-padding	boolean	false	否	是否去掉 iOS 下的默认内边距	2.10.0
confirm-type	string	return	否	设置键盘右下角按钮的文字。 send：按钮文字为“发送”； search：按钮文字为“搜索”； next：按钮文字为“下一个”； go：按钮文字为“前往”； done：按钮文字为“完成”； return：按钮文字为“换行”	2.13.0
confirm-hold	boolean	false	否	单击键盘右下角按钮时是否保持键盘不收起	2.16.0
bindfocus	eventhandle		否	输入框聚焦时触发，event.detail = { value, height }，height 为键盘高度，在基础库 1.9.90 起支持	1.0.0
bindblur	eventhandle		否	输入框失去焦点时触发，event.detail = {value, cursor}	1.0.0
bindlinechange	eventhandle		否	输入框行数变化时调用，event.detail = {height: 0, heightRpx: 0, lineCount: 0}	1.0.0
bindinput	eventhandle		否	当键盘输入时，触发 input 事件，event.detail = {value, cursor, keyCode}，keyCode 为键值，目前工具还不支持返回 keyCode 参数。bindinput 处理函数的返回值并不会反映到 textarea 上	1.0.0
bindconfirm	eventhandle		否	单击“完成”按钮时触发 confirm 事件，event.detail= {value: value}	1.0.0
bindkeyboardheightchange	eventhandle		否	键盘高度发生变化时触发此事件，event.detail = {height: height, duration: duration}	2.7.0

WXML 程序代码如下。

```
<form bindsubmit='fun_bindsubmit' bindreset='fun_bindreset'>
  <view>textarea
    <textarea bindblur="bindTextAreaBlur" auto-height placeholder="自动变高" />
    <textarea placeholder="placeholder颜色是红色的" placeholder-style="color:red;"  />
```

```
    </view>

    <button type="primary" size="mini" form-type="submit">submit</button>
    <button type="primary" size="mini" form-type="reset">reset</button>
  </form>
```

程序编译后的效果如图 4-29 所示。

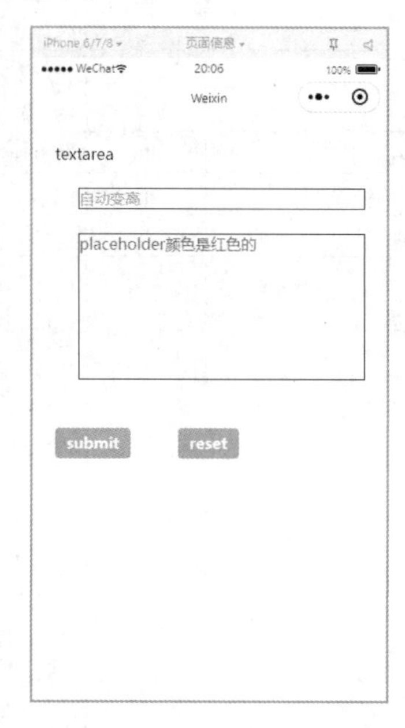

图 4-29

4.4.4 checkbox 组件

checkbox 是多选项目组件，其属性如表 4-9 所示。

表 4-9

属　性	类　型	默 认 值	必　填	说　明	最 低 版 本
value	string		否	checkbox 标识，选中时触发 checkbox-group 的 change 事件，并携带 checkbox 的 value	1.0.0
disabled	boolean	false	否	是否禁用	1.0.0
checked	boolean	false	否	当前是否选中，可用来设置默认选中	1.0.0
color	string	#09BB07	否	checkbox 的颜色，同 CSS 的 color	1.0.0

WXML 程序代码如下。

```
<form bindsubmit='fun_bindsubmit' bindreset='fun_bindreset'>

  <view>
    name='weixin'
    <view style="margin:0">
      <checkbox-group name="weixin">
        <label>
          <checkbox value="微信小程序" checked="true" />小程序</label>
        <label>
```

```
        <checkbox value="微信公众号" checked="true" />公众号</label>
      <label>
        <checkbox value="企业微信" checked="true" />企业微信</label>
      <label>
        <checkbox value="微信支付" checked="true" />支付</label>
    </checkbox-group>
  </view>
</view>

<view>
  name='jiubao'
  <view style="margin:0">
    <checkbox-group name="jiubao">
      <label>
        <checkbox value="唐诗" checked="true" />古诗</label>
      <label>
        <checkbox value="宋词" checked="true" />古词</label>
      <label>
        <checkbox value="元曲" checked="true" />古曲</label>
    </checkbox-group>
  </view>
</view>

<button type="primary" size="mini" form-type="submit">submit</button>
<button type="primary" size="mini" form-type="reset">reset</button>

</form>
```

程序编译后的效果如图 4-30 所示。

图 4-30

checkbox-group 是 checkbox 的 group，微信小程序中 checkbox 的 name 需要由 checkbox-group 进行定义。

4.4.5 switch 组件

switch 是开关选择器组件，其属性如表 4-10 所示。

表 4-10

属　性	类　型	默认值	必　填	说　明	最低版本
checked	boolean	false	否	是否选中	1.0.0
disabled	boolean	false	否	是否禁用	1.0.0
type	string	switch	否	样式，有效值包括 switch 和 checkbox	1.0.0
color	string	#04BE02	否	switch 的颜色，同 CSS 的 color	1.0.0
bindchange	eventhandle		否	checked 改变时触发 change 事件，event.detail ={ value}	1.0.0

WXML 程序代码如下。

```
<form bindsubmit='fun_bindsubmit' bindreset='fun_bindreset'>

  <view><label>
     <switch value="韩" checked="true" />韩</label></view>
  <view><label>
     <switch value="赵" checked="{{false}}" />赵</label></view>
  <view><label>
     <switch value="魏" checked="true" />魏</label></view>
  <view><label>
     <switch value="楚" checked="true" />楚</label></view>
  <view><label>
     <switch value="齐" checked="true" type="checkbox"/>齐</label></view>
  <view><label>
     <switch value="秦" checked="{{false}}" type="checkbox"/>秦</label></view>

  <button type="primary" size="mini" form-type="submit">submit</button>
  <button type="primary" size="mini" form-type="reset">reset</button>

</form>
```

程序编译后的效果如图 4-31 所示。

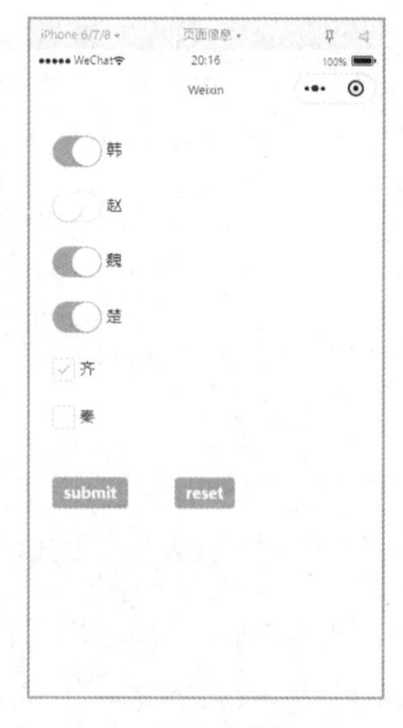

图 4-31

4.4.6 radio 组件

radio 是单选项目组件，其属性如表 4-11 所示。

表 4-11

属　性	类　型	默认值	必　填	说　明	最低版本
value	string		否	radio 标识。当该 radio 选中时，radio-group 的 change 事件会携带 radio 的 value	1.0.0
checked	boolean	false	否	当前是否选中	1.0.0
disabled	boolean	false	否	是否禁用	1.0.0
color	string	#09BB07	否	radio 的颜色，同 CSS 的 color	1.0.0

WXML 程序代码如下。

```
<form bindsubmit='fun_bindsubmit' bindreset='fun_bindreset'>

  <view>
    name='weixin'
    <view style="margin:0">
      <radio-group name="weixin">
        <label>
          <radio value="微信小程序" checked="true" />小程序</label>
        <label>
          <radio value="微信公众号"/>公众号</label>
        <label>
          <radio value="企业微信"/>企业微信</label>
        <label>
          <radio value="微信支付"/>支付</label>
      </radio-group>
    </view>
  </view>

  <view>
    name='jiubao'
    <view style="margin:0">
      <radio-group name="jiubao">
        <label>
          <radio value="唐诗" checked="true" />古诗</label>
        <label>
          <radio value="宋词"/>古词</label>
        <label>
          <radio value="元曲"/>古曲</label>
      </radio-group>
    </view>
  </view>

  <button type="primary" size="mini" form-type="submit">submit</button>
  <button type="primary" size="mini" form-type="reset">reset</button>

</form>
```

程序编译后效果如图 4-32 所示。

图 4-32

radio-group 是 radio 的 group。微信小程序中 radio 的 name 是由 radio-group 定义的。

4.4.7 keyboard-accessory 组件

keyboard-accessory 组件用于设置 input / textarea 组件聚焦时键盘上方的 cover-view / cover-image 工具栏视图。

WXML 程序代码如下。

```
<form bindsubmit='fun_bindsubmit' bindreset='fun_bindreset'>

  <view>
    <input hold-keyboard>
      <keyboard-accessory style="display: flex;height: 50px;">
        <cover-view style="flex: 1; background: black ; color: white; ">1</cover-view>
        <cover-view style="flex: 1; background: blue  ; color: white;">2</cover-view>
      </keyboard-accessory>
    </input>

    <textarea hold-keyboard>
      <keyboard-accessory style="display: flex;height: 50px;">
        <cover-view style="flex: 1; background: black ; color: white; ">1</cover-view>
        <cover-view style="flex: 1; background: red  ; color: white;">2</cover-view>
      </keyboard-accessory>
    </textarea>
  </view>

  <button type="primary" size="mini" form-type="submit">submit</button>
  <button type="primary" size="mini" form-type="reset">reset</button>

</form>
```

程序编译后的初始效果如图 4-33 所示。当 input 组件获取焦点时，键盘上方 cover-view/cover-image 工具栏视图显示蓝色背景，效果如图 4-34 所示。当 textarea 组件获取焦点时，键盘上方 cover-view/cover-image 工具栏视图显示红色背景，效果如图 4-35 所示。

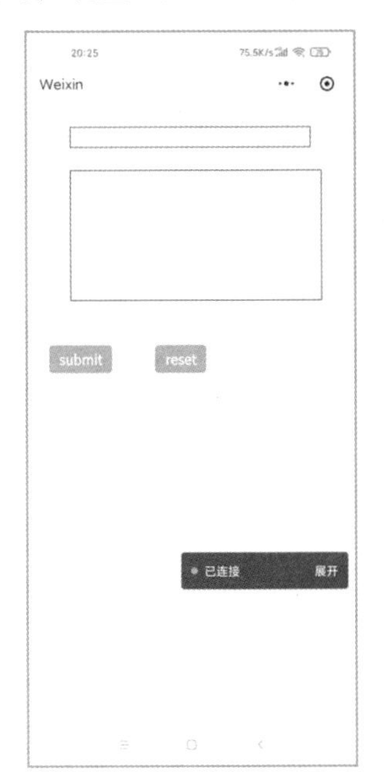

图 4-33

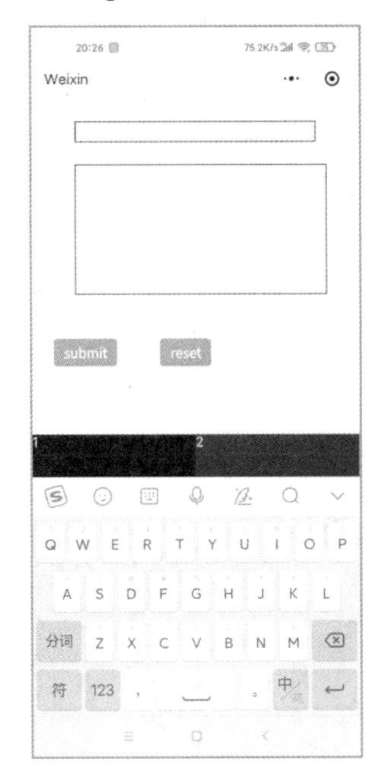

图 4-34

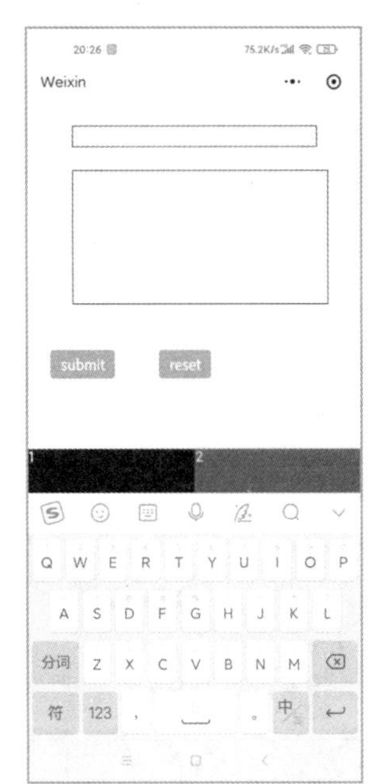

图 4-35

4.4.8　label 组件

label 组件使用 for 属性找到对应的 id，或者将组件放在该标签下，单击时触发对应的组件。for 属性的优先级高于内部组件，当内部有多个组件时，默认触发第一个组件。目前可以绑定的组件有 button 组件、checkbox 组件、radio 组件和 switch 组件。

lable 组件的属性如表 4-12 所示。

表 4-12

属　性	类　型	默 认 值	必　填	说　明	最 低 版 本
for	string		否	绑定控件的 id	1.0.0

WXML 程序代码如下。

```
<form bindsubmit='fun_bindsubmit' bindreset='fun_bindreset'>

  <view>
    <radio-group name="jiubao">
      <label>
        <radio value="唐诗" checked="true" />古诗</label>
      <label>
        <radio value="宋词" />古词</label>
      <label>
        <radio value="元曲" />古曲</label>
    </radio-group>
  </view>

  <view>
    <radio-group name="jiubao">
      <label for='radio_gs'>古诗</label>
      <radio id='radio_gs' value="唐诗" checked="true" />
      <label for='radio_gc'>古词</label>
      <radio id='radio_gc' value="宋词" />
```

```
      <label for='radio_gq'>古曲</label>
      <radio id='radio_gq' value="元曲" />
    </radio-group>
  </view>

  <button type="primary" size="mini" form-type="submit">submit</button>
  <button type="primary" size="mini" form-type="reset">reset</button>
</form>
```

程序编译后效果如图 4-36 所示。

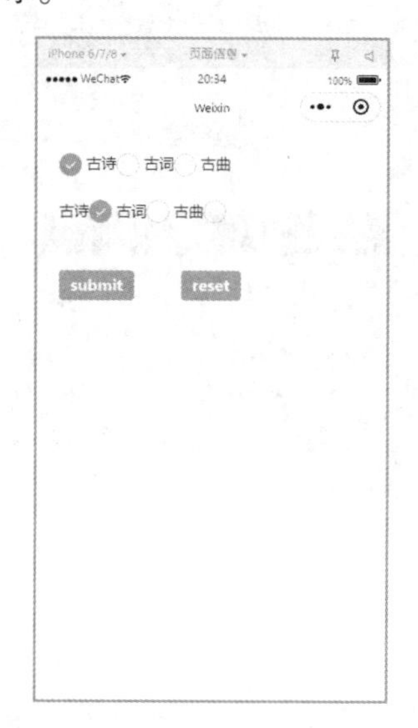

图 4-36

4.4.9　slider 组件

slider 是滑动选择器组件，其属性如表 4-13 所示。

表 4-13

属　性	类　型	默 认 值	必　填	说　明	最 低 版 本
min	number	0	否	最小值	1.0.0
max	number	100	否	最大值	1.0.0
step	number	1	否	步长，取值必须大于 0，并且可被(max - min)整除	1.0.0
disabled	boolean	false	否	是否禁用	1.0.0
value	number	0	否	当前取值	1.0.0
color	color	#e9e9e9	否	背景条的颜色（请使用 backgroundColor）	1.0.0
selected-color	color	#1aad19	否	已选择的颜色（请使用 activeColor）	1.0.0
activeColor	color	#1aad19	否	已选择的颜色	1.0.0
backgroundColor	color	#e9e9e9	否	背景条的颜色	1.0.0
block-size	number	28	否	滑块的大小，取值范围为 12 ～ 28	1.9.0
block-color	color	#ffffff	否	滑块的颜色	1.9.0
show-value	boolean	false	否	是否显示当前 value	1.0.0
bindchange	eventhandle		否	完成一次拖动后触发的事件，event.detail={value}	1.0.0
bindchanging	eventhandle		否	拖动过程中触发的事件，event.detail = {value}	1.7.0

WXML 程序代码如下。

```
<form bindsubmit='fun_bindsubmit' bindreset='fun_bindreset'>
  <view>
    slider
    <view>
      <slider />
    </view>
  </view>

  <view class="section section_gap">
    设置 step
    <view>
      <slider step="15" />
    </view>
  </view>

  <view class="section section_gap">
    显示当前 value
    <view>
      <slider show-value />
    </view>
  </view>

  <view class="section section_gap">
    设置最小/最大值
    <view>
      <slider min="50" max="200" show-value />
    </view>
  </view>

  <button type="primary" size="mini" form-type="submit">submit</button>
  <button type="primary" size="mini" form-type="reset">reset</button>
</form>
```

程序编译后效果如图 4-37 所示。

图 4-37

4.5 导航组件

navigator 导航组件的主要作用是跳转导航，导航组件属性如表 4-14 所示。

表 4-14

组 件	属 性	类 型	默 认 值	必 填	说 明	最 低 版 本
navigator	target	string	self	否	在哪个目标上发生跳转，默认为当前小程序	2.0.7
	url	string		否	当前小程序内的跳转链接	1.0.0
	open-type	string	navigate	否	跳转方式	1.0.0
	delta	number	1	否	当 open-type 为'navigateBack' 时有效，表示回退的层数	1.0.0
	app-id	string		否	当 target="miniProgram"时有效，表示要打开的小程序 appId	2.0.7
	path	string		否	当 target="miniProgram"时有效，表示打开的页面路径，如果为空则打开首页	2.0.7
	extra-data	object		否	当 target="miniProgram"时有效，表示需要传递给目标小程序的数据，目标小程序可在 App.onLaunch()和 App.onShow() 中获取到这份数据	2.0.7
	version	string	release	否	当 target="miniProgram"时有效，表示要打开的小程序版本	2.0.7
	hover-class	string	navigator-hover	否	指定单击时的样式类，当 hover-class="none"时没有单击态效果	1.0.0
	hover-stop-propagation	boolean	false	否	指定是否阻止本节点的祖先节点出现单击态	1.5.0
	hover-start-time	number	50	否	按住后多久出现单击态，单位为毫秒	1.0.0
	hover-stay-time	number	600	否	手指松开后单击态保留时间，单位为毫秒	1.0.0
	bindsuccess	string		否	当 target="miniProgram"时有效，表示跳转小程序成功	2.0.7
	bindfail	string		否	当 target="miniProgram"时有效，表示跳转小程序失败	2.0.7
	bindcomplete	string		否	当 target="miniProgram"时有效，表示跳转小程序完成	2.0.7
target	self				当前小程序	
	miniProgram				其他小程序	
open-type	navigate				对应 wx.navigateTo 或 wx.navigateToMiniProgram 的功能	
	redirect				对应 wx.redirectTo 的功能	
	switchTab				对应 wx.switchTab 的功能	
	reLaunch				对应 wx.reLaunch 的功能	
	navigateBack				对应 wx.navigateBack 的功能	
	exit				退出小程序，当 target="miniProgram"时生效	

续表

组　件	属　性	类　型	默 认 值	必　填	说　明	最 低 版 本
version	develop				开发版	
	trial				体验版	
	release				正式版，此参数仅在当前小程序为开发版或体验版时有效；如果当前小程序是正式版，则打开的小程序必定是正式版	

下面来看一下 navigator 的示例程序。其中，index 是主页面，navigate 是跳转的新页面，redirect 是在主页面中打开的页面。

1. index 相关程序

（1）index 没有 JS 程序

（2）WXML 程序代码如下。

```
<view class="btn-area">
  <navigator url="/pages/navigate/navigate?title=navigate" open-type="navigate"
hover-class="navigator-hover">跳转到新页面</navigator>
  <navigator url="/pages/redirect/redirect?title=redirect" open-type="redirect"
hover-class="navigator-hover1">在当前页打开</navigator>
</view>
```

（3）WXSS 程序代码如下。

```
/** 修改默认的 navigator 单击态 **/
.navigator-hover {
  color:blue;
}
/** 自定义其他单击态样式类 **/
.navigator-hover1 {
  color:red;
}

navigator{
  margin: 40rpx;
}
```

2. navigate 相关程序

（1）JS 程序代码如下。

```
// pages/navigate/navigate.js
Page({

  data: {
    str:''
  },

  onLoad: function (options) {
    this.setData({str:options.title});
  },

})
```

（2）WXML 程序代码如下。

```
<!--pages/navigate/navigate.wxml-->
```

```
<view>
  <view>pages/navigate/navigate.wxml</view>
  参数是: {{str}}
</view>
```

（3）WXSS 程序代码如下。

```
/* pages/navigate/navigate.wxss */
view{
  margin: 40rpx;
}
```

3. redirect 相关程序

（1）JS 程序代码如下。

```
// pages/redirect/redirect.js
Page({

  data: {
    str:''
  },

  onLoad: function (options) {
    this.setData({str:options.title});
  },

})
```

（2）WXML 程序代码如下。

```
<!--pages/redirect/redirect.wxml-->
<view>
  <view>pages/redirect/redirect.wxml</view>
  参数是: {{str}}
</view>
```

（3）WXSS 程序代码如下。

```
/* pages/redirect/redirect.wxss */
view{
  margin: 40rpx;
}
```

需要特别注意的是，app.json 需要配置 navigate 和 redirect，代码如下。

```
{
  "pages": [
    "pages/index/index",
    "pages/navigate/navigate",
    "pages/redirect/redirect"
  ],
  "window": {
    "backgroundTextStyle": "light",
    "navigationBarBackgroundColor": "#fff",
    "navigationBarTitleText": "Weixin",
    "navigationBarTextStyle": "black"
  },
  "style": "v2",
  "sitemapLocation": "sitemap.json"
}
```

程序编译后初始效果如图 4-38 所示。单击“跳转到新页面”按钮后，此处文字颜色由黑色变为蓝色，如图 4-39 所示。

使用 navigate 跳转页面时，微信小程序将保留当前页面（tabBar 页面除外），如图 4-40 所示。注意观察左上角的图标。使用 wx.navigateBack 可以返回原页面。微信小程序中页面栈最多有 10 层。

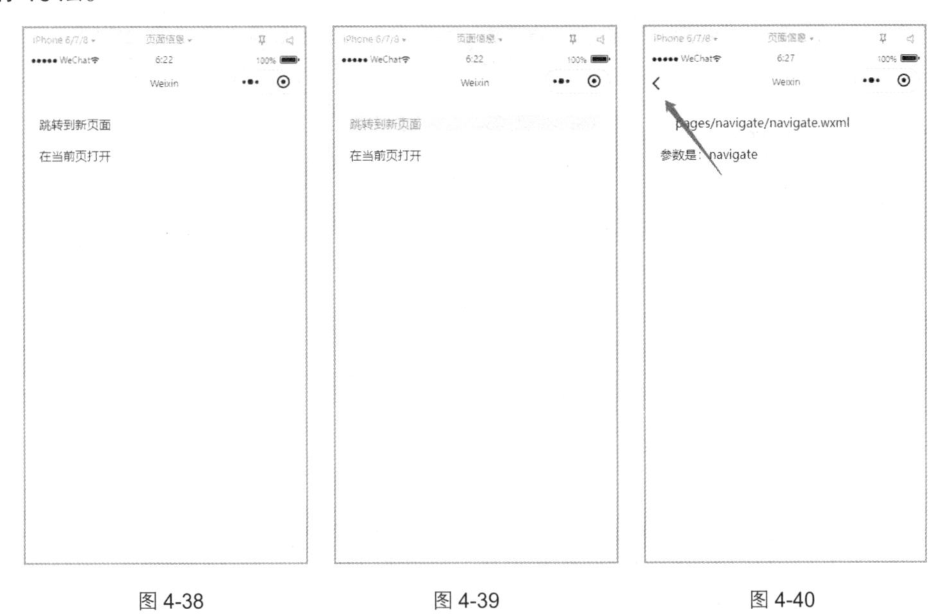

图 4-38　　图 4-39　　图 4-40

返回 index 页面，单击“在当前页打开”按钮，此时文字颜色由黑色变为红色，如图 4-41 所示。

与使用 navigate 跳转页面不同，使用 redirect 跳转页面时，微信小程序会关闭当前页面，如图 4-42 所示。注意左上角图标。

图 4-41

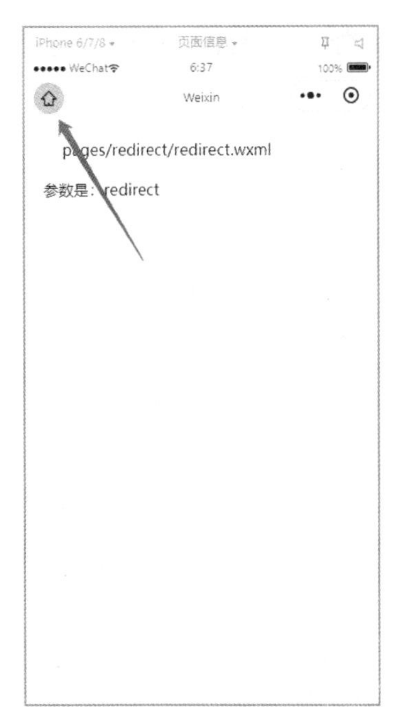

图 4-42

通过以上示例,读者可以清晰地认识到 navigate 与 redirect 的区别。若想了解更多 navigator 的其他属性，读者可以参考腾讯官方文档。

4.6 媒体组件

媒体组件主要包括 audio 组件、camera 组件、image 组件、video 组件，下面分别进行介绍。

4.6.1 audio 组件

audio 组件可以实现与音频有关的功能。但从微信 1.6.0 版本开始，该组件不再被支持和维护，建议读者使用功能更强的 wx.createInnerAudioContext 接口。

注意：

wx.createInnerAudioContext 接口相关程序的注意事项较多，应用时应谨慎。

下面以自定义音频组件为例，说明相关的知识点。

（1）JS 程序代码如下。

```
let innerAudioContext;
Page({
  data: {
    max:0,
    value:0,
    type:'',
  },
  onReady:function(){
    this.create();
  },
  create: function () {
    this.setData({type:'create'})
    console.log('构建');
    innerAudioContext = wx.createInnerAudioContext()
    innerAudioContext.src = '少年.mp3'
    innerAudioContext.onPlay(() => {
      console.log('开始播放')
    })
    innerAudioContext.onError((res) => {
      console.log(res.errMsg)
      console.log(res.errCode)
    })
    innerAudioContext.onTimeUpdate(() => {
      this.setData({max:innerAudioContext.duration});
      this.setData({value:innerAudioContext.currentTime});
      console.log('总时长: ', innerAudioContext.duration, '当前播放进度: ',
innerAudioContext.currentTime);
    });
    innerAudioContext.onPause(() => {
      console.log('暂停')
    })
```

```
    innerAudioContext.onStop(() => {
      console.log('停止播放')
    })
    innerAudioContext.onSeeked(()=>{
      if('stop'!=this.data.type){
        innerAudioContext.play()
      }
    })
    innerAudioContext.onCanplay(()=>{
      if('stop'!=this.data.type){
        innerAudioContext.play()
      }
    })
  },
  play: function () {
    this.setData({type:'play'})
    innerAudioContext.play()
  },
  pause: function () {
    this.setData({type:'pause'})
    innerAudioContext.pause()
  },
  seek: function () {
    this.setData({type:'seek'})
    innerAudioContext.pause()
    innerAudioContext.seek(10)
  },
  stop: function () {
    this.setData({type:'stop'})
    innerAudioContext.stop()
  },
  bindchange:function(event){
    innerAudioContext.pause()
    innerAudioContext.seek(event.detail.value)
  }
})
```

（2）WXML 程序代码如下。

```
<button type="primary" bindtap="create">构建</button>
<button type="primary" bindtap="play">播放</button>
<button type="primary" bindtap="pause">暂停</button>
<button type="primary" bindtap="seek">跳转到第 10 秒</button>
<button type="primary" bindtap="stop">停止</button>
<slider max="{{max}}" value="{{value}}" min="0" show- show-value="{{true}}"
bindchange="bindchange"
  backgroundColor="#000" block-color="#f00" activeColor="#00f" />
```

（3）WXSS 程序代码如下。

```
button{
  margin: 40rpx;
}
```

程序编译后效果如图 4-43 所示。

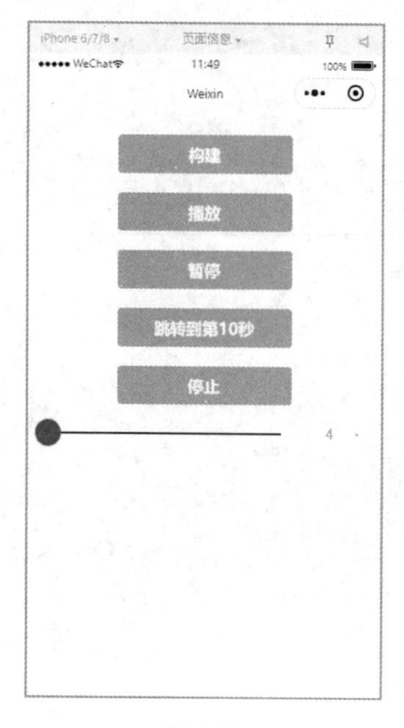

图 4-43

以下变量用于逻辑控制。其中，max 控制 slider 的最大值，value 控制 slider 的当前值，type 用于区分时间类型。

```
data: {
  max:0,
  value:0,
  type:'',
},
```

“构建”按钮用于实现初始化 innerAudioContext，以及向 innerAudioContext 注册相关事件。需要监听的事件有 onPlay、onError、onTimeUpdate、onPause、onStop、onSeeked、onCanplay。相关程序代码如下。

```
create: function () {
    this.setData({type:'create'})
    console.log('构建');
    innerAudioContext = wx.createInnerAudioContext()
    innerAudioContext.src = '少年.mp3'
    innerAudioContext.onPlay(() => {
      console.log('开始播放')
    })
    innerAudioContext.onError((res) => {
      console.log(res.errMsg)
      console.log(res.errCode)
    })
    innerAudioContext.onTimeUpdate(() => {
      this.setData({max:innerAudioContext.duration});
      this.setData({value:innerAudioContext.currentTime});
      console.log('总时长: ', innerAudioContext.duration, '当前播放进度: ',
innerAudioContext.currentTime);
    });
    innerAudioContext.onPause(() => {
```

```
      console.log('暂停')
    })
    innerAudioContext.onStop(() => {
      console.log('停止播放')
    })
    innerAudioContext.onSeeked(()=>{
      if('stop'!=this.data.type){
        innerAudioContext.play()
      }
    })
    innerAudioContext.onCanplay(()=>{
      if('stop'!=this.data.type){
        innerAudioContext.play()
      }
    })
  },
```

“播放”按钮可实现音频播放，相关程序代码如下。

```
  play: function () {
    this.setData({type:'play'})
    innerAudioContext.play()
  },
```

“暂停”按钮可实现音频暂停播放，相关程序代码如下。

```
  pause: function () {
    this.setData({type:'pause'})
    innerAudioContext.pause()
  },
```

“跳转到第 10 秒”按钮可实现跳转到当前音频的第 10 秒，相关程序代码如下。

```
  seek: function () {
    this.setData({type:'seek'})
    innerAudioContext.pause()
    innerAudioContext.seek(10)
  },
```

“停止”按钮可实现音频停止播放，相关程序代码如下。

```
  stop: function () {
    this.setData({type:'stop'})
    innerAudioContext.stop()
  },
```

bindchange 可绑定 slider 组件的 bindchange 事件，通过 slider 控制音频播放，相关程序代码如下。

```
  bindchange:function(event){
    innerAudioContext.pause()
    innerAudioContext.seek(event.detail.value)
  }
```

微信小程序音频处理拥有多个版本，到本书截稿为止，腾讯推荐的是 InnerAudioContext 方式。InnerAudioContext 通过 wx.createInnerAudioContext 接口获取。

注意：

音频播放过程中可能被系统中断，可通过 wx.onAudioInterruptionBegin、wx.onAudioInterruptionEnd 事件来处理这种情况。

InnerAudioContext 的常用属性如下。

- string src：音频资源的地址，用于直接播放。从 2.2.3 版本开始支持云文件 ID。
- number startTime：开始播放的位置（单位：s），默认为 0。
- boolean autoplay：是否自动开始播放，默认为 false。
- boolean loop：是否循环播放，默认为 false。
- boolean obeyMuteSwitch：是否遵循系统静音开关，默认为 true。当此参数为 false 时，即使用户打开了静音开关，也能继续发出声音。从 2.3.0 版本开始此参数不生效，使用 wx.setInnerAudioOption 接口统一设置。
- number volume：音量，范围为 0~1，默认为 1。基础库 1.9.90 开始支持，低版本需做兼容处理。
- number playbackRate：播放速度，范围为 0.5~2.0，默认为 1（Android 需要 6 及以上版本）。基础库 2.11.0 开始支持，低版本需做兼容处理。
- number duration：当前音频的长度（单位：s）。只有在当前有合法的 src 时返回（只读）。
- number currentTime：当前音频的播放位置（单位：s）。只有在当前有合法的 src 时返回，时间保留小数点后 6 位（只读）。
- boolean paused：当前是否是暂停或停止状态（只读）。
- number buffered：音频缓冲的时间点，仅保证当前播放时间点到此时间点内容已缓冲（只读）。

InnerAudioContext 方法如下。

- InnerAudioContext.play()：播放。
- InnerAudioContext.pause()：暂停。暂停后的音频再播放时会从上次暂停处开始播放。
- InnerAudioContext.stop()：停止。停止后的音频再播放时会从头开始播放。
- InnerAudioContext.seek(number position)：跳转到指定位置。
- InnerAudioContext.destroy()：销毁当前实例。
- InnerAudioContext.onCanplay(function callback)：监听音频进入可以播放状态的事件。但不保证后面可以流畅播放。
- InnerAudioContext.offCanplay(function callback)：取消监听音频进入可以播放状态的事件。
- InnerAudioContext.onPlay(function callback)：监听音频播放事件。
- InnerAudioContext.offPlay(function callback)：取消监听音频播放事件。
- InnerAudioContext.onPause(function callback)：监听音频暂停事件。
- InnerAudioContext.offPause(function callback)：取消监听音频暂停事件。
- InnerAudioContext.onStop(function callback)：监听音频停止事件。
- InnerAudioContext.offStop(function callback)：取消监听音频停止事件。
- InnerAudioContext.onEnded(function callback)：监听音频自然播放至结束的事件。
- InnerAudioContext.offEnded(function callback)：取消监听音频自然播放至结束的事件。
- InnerAudioContext.onTimeUpdate(function callback)：监听音频播放进度更新事件。
- InnerAudioContext.offTimeUpdate(function callback)：取消监听音频播放进度更新事件。
- InnerAudioContext.onError(function callback)：监听音频播放错误事件。

◆ InnerAudioContext.offError(function callback)：取消监听音频播放错误事件。
◆ InnerAudioContext.onWaiting(function callback)：监听音频加载中事件。当音频因为数据不足需要停下来加载时会触发。
◆ InnerAudioContext.offWaiting(function callback)：取消监听音频加载中事件。
◆ InnerAudioContext.onSeeking(function callback)：监听音频进行跳转操作的事件。
◆ InnerAudioContext.offSeeking(function callback)：取消监听音频进行跳转操作的事件。
◆ InnerAudioContext.onSeeked(function callback)：监听音频完成跳转操作的事件。
◆ InnerAudioContext.offSeeked(function callback)：取消监听音频完成跳转操作的事件。

InnerAudioContext 支持格式如表 4-15 所示。

表 4-15

格　　式	iOS	Android	格　　式	iOS	Android
flac	×	√	wav	√	√
m4a	√	√	mp3	√	√
ogg	×	√	mp4	×	√
ape	×	√	aac	√	√
amr	×	√	aiff	√	×
wma	×	√	caf	√	×

4.6.2　camera 组件

camera 组件可实现系统相机的有关功能。

注意：

要使用相机实现扫描二维码功能，微信客户端需升级至 6.7.3 版本，同时需用户授权 scope.camera。从 2.10.0 版本起，initdone 事件返回 maxZoom，最大变焦范围，相关接口：CameraContext.setZoom。

camera 组件的相关属性如表 4-16 所示。

表 4-16

属　　性	类　　型	默 认 值	必　　填	说　　明	最 低 版 本
mode	string	normal	否	应用模式，只在初始化时有效，不能动态变更。normal：相机模式；scanCode：扫码模式	2.1.0
resolution	string	medium	否	分辨率，不支持动态修改。取值包括 low（低）、medium（中）和 high（高）	2.10.0
device-position	string	back	否	摄像头朝向，取值包括 front（前置）和 back（后置）	1.0.0
flash	string	auto	否	闪光灯，取值包括 auto（自动）、on（打开）、off（关闭）。2.8.0 版本后取值还包括 torch（常亮）	1.0.0
frame-size	string	medium	否	指定期望的相机帧数据尺寸，取值包括 small、medium 和 large	2.7.0
bindstop	eventhandle		否	摄像头在非正常终止时触发，如退出后台等情况	1.0.0
binderror	eventhandle		否	用户不允许使用摄像头时触发	1.0.0
bindinitdone	eventhandle		否	相机初始化完成时触发，e.detail = {maxZoom}	2.7.0
bindscancode	eventhandle		否	在扫码识别成功时触发，仅在 mode="scanCode" 时生效	2.1.0

（1）JS 程序代码如下。

```
Page({
  data:{
    src:''
  },
  takePhoto() {
    const ctx = wx.createCameraContext()
    ctx.takePhoto({
      quality: 'high',
      success: (res) => {
        this.setData({
          src: res.tempImagePath
        })
      }
    })
  },
  error(e) {
    console.log(e.detail)
  }
})
```

（2）WXMLl 程序代码如下。

```
<camera device-position="back" flash="off" binderror="error" style="width: 100%;
height: 300px;"></camera>
<button type="primary" bindtap="takePhoto">拍照</button>
<view>预览</view>
<image mode="widthFix" src="{{src}}"></image>
```

（3）WXSS 程序代码如下。

```
button,view{
  margin: 40rpx;
}
```

程序编译后效果如图 4-44 所示。

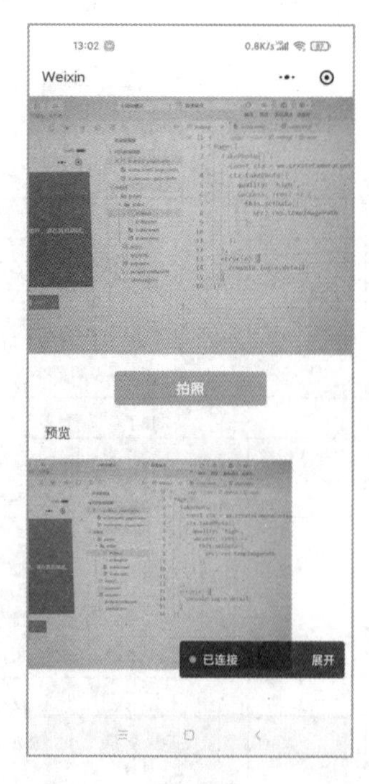

图 4-44

要想使用 camera 实现扫码功能，可以参照下文编写代码。

（1）JS 程序代码如下。

```
Page({
  data:{
    str:'-'
  },
  cancode(e){
      this.setData({str:e.detail.result});
  },
  error(e) {
    console.log(e.detail)
  }
})
```

（2）WXML 程序代码如下。

```
<camera mode='scanCode' device-position="back" flash="off" bindscancode='cancode'
binderror="error" style="width: 100%; height: 300px;"></camera>
<view>结果</view>
<view>{{str}}</view>
```

（3）WXSS 程序代码如下。

```
button,view{
  margin: 40rpx;
}
```

成功编译微信小程序后，扫描一个二维码，效果如图 4-45 所示。

图 4-45

4.6.3 image 组件

image 组件可实现图片功能。支持 JPG、PNG、SVG、WEBP、GIF 等格式，2.3.0 版本之后还支持云文件 ID。

image 组件的相关属性如表 4-17 所示。

表 4-17

属　　性	类　　型	默 认 值	必　　填	说　　明	最 低 版 本
src	string		否	图片资源地址	1.0.0
mode	string	scaleToFill	否	图片裁剪、缩放的模式	1.0.0
webp	boolean	false	否	默认不解析 webP 格式，只支持网络资源	2.9.0
lazy-load	boolean	false	否	图片懒加载，在即将进入一定范围（上下三屏）时才开始加载	1.5.0
show-menu-by-longpress	boolean	false	否	开启长按图片显示识别小程序码菜单	2.7.0
binderror	eventhandle		否	当错误发生时触发，event.detail = {errMsg}	1.0.0
bindload	eventhandle		否	当图片载入完毕时触发，event.detail = {height, width}	1.0.0

mode 属性的合法值如表 4-18 所示。

表 4-18

值	说　　明
scaleToFill	缩放模式，不保持纵横比缩放图片，使图片的宽高完全拉伸至填满 image 元素
aspectFit	缩放模式，保持纵横比缩放图片，使图片的长边能完全显示出来。也就是说，可以完整地将图片显示出来
aspectFill	缩放模式，保持纵横比缩放图片，只保证图片的短边能完全显示出来。也就是说，图片通常只在水平或垂直方向是完整的，另一个方向将会发生截取
widthFix	缩放模式，宽度不变，高度自动变化，保持原图宽高比不变
heightFix	缩放模式，高度不变，宽度自动变化，保持原图宽高比不变
top	裁剪模式，不缩放图片，只显示图片的顶部区域
bottom	裁剪模式，不缩放图片，只显示图片的底部区域
center	裁剪模式，不缩放图片，只显示图片的中间区域
left	裁剪模式，不缩放图片，只显示图片的左边区域
right	裁剪模式，不缩放图片，只显示图片的右边区域
top left	裁剪模式，不缩放图片，只显示图片的左上边区域
top right	裁剪模式，不缩放图片，只显示图片的右上边区域
bottom left	裁剪模式，不缩放图片，只显示图片的左下边区域
bottom right	裁剪模式，不缩放图片，只显示图片的右下边区域

（1）JS 程序代码如下。

```
Page({
  data: {
    array: [{
      mode: 'scaleToFill',
      text: 'scaleToFill: 不保持纵横比缩放图片，使图片完全适应'
    }, {
      mode: 'aspectFit',
      text: 'aspectFit: 保持纵横比缩放图片，使图片的长边能完全显示出来'
    }, {
      mode: 'aspectFill',
      text: 'aspectFill: 保持纵横比缩放图片，只保证图片的短边能完全显示出来'
    }, {
      mode: 'top',
```

```
      text: 'top: 不缩放图片，只显示图片的顶部区域'
    }, {
      mode: 'bottom',
      text: 'bottom: 不缩放图片，只显示图片的底部区域'
    }, {
      mode: 'center',
      text: 'center: 不缩放图片，只显示图片的中间区域'
    }, {
      mode: 'left',
      text: 'left: 不缩放图片，只显示图片的左边区域'
    }, {
      mode: 'right',
      text: 'right: 不缩放图片，只显示图片的右边区域'
    }, {
      mode: 'top left',
      text: 'top left: 不缩放图片，只显示图片的左上边区域'
    }, {
      mode: 'top right',
      text: 'top right: 不缩放图片，只显示图片的右上边区域'
    }, {
      mode: 'bottom left',
      text: 'bottom left: 不缩放图片，只显示图片的左下边区域'
    }, {
      mode: 'bottom right',
      text: 'bottom right: 不缩放图片，只显示图片的右下边区域'
    }],
    src: '../../直角坐标.jpg'
  },
  imageError: function(e) {
    console.log('image3 发生 error 事件，携带值为', e.detail.errMsg)
  }
})
```

（2）WXML 程序代码如下。

```
<view>
  <view>
    <view wx:for="{{array}}" wx:for-item="item">
      <view>{{item.text}}</view>
      <view>
        <image style="width: 250px; height: 250px; background-color: #f00;"
mode="{{item.mode}}" src="{{src}}"></image>
      </view>
    </view>
  </view>
</view>
```

（3）WXSS 程序代码如下。

```
image{
  border: 1px blue solid;
}
```

注意：

该示例微信小程序 image 的宽、高都是 250 px，图片的宽为 300 px、高为 250 px。

为了能看清楚图片效果，示例程序为 image 增加了 border，同时定义 image 的 background-color 为#f00。程序编译后效果如图 4-46 所示。

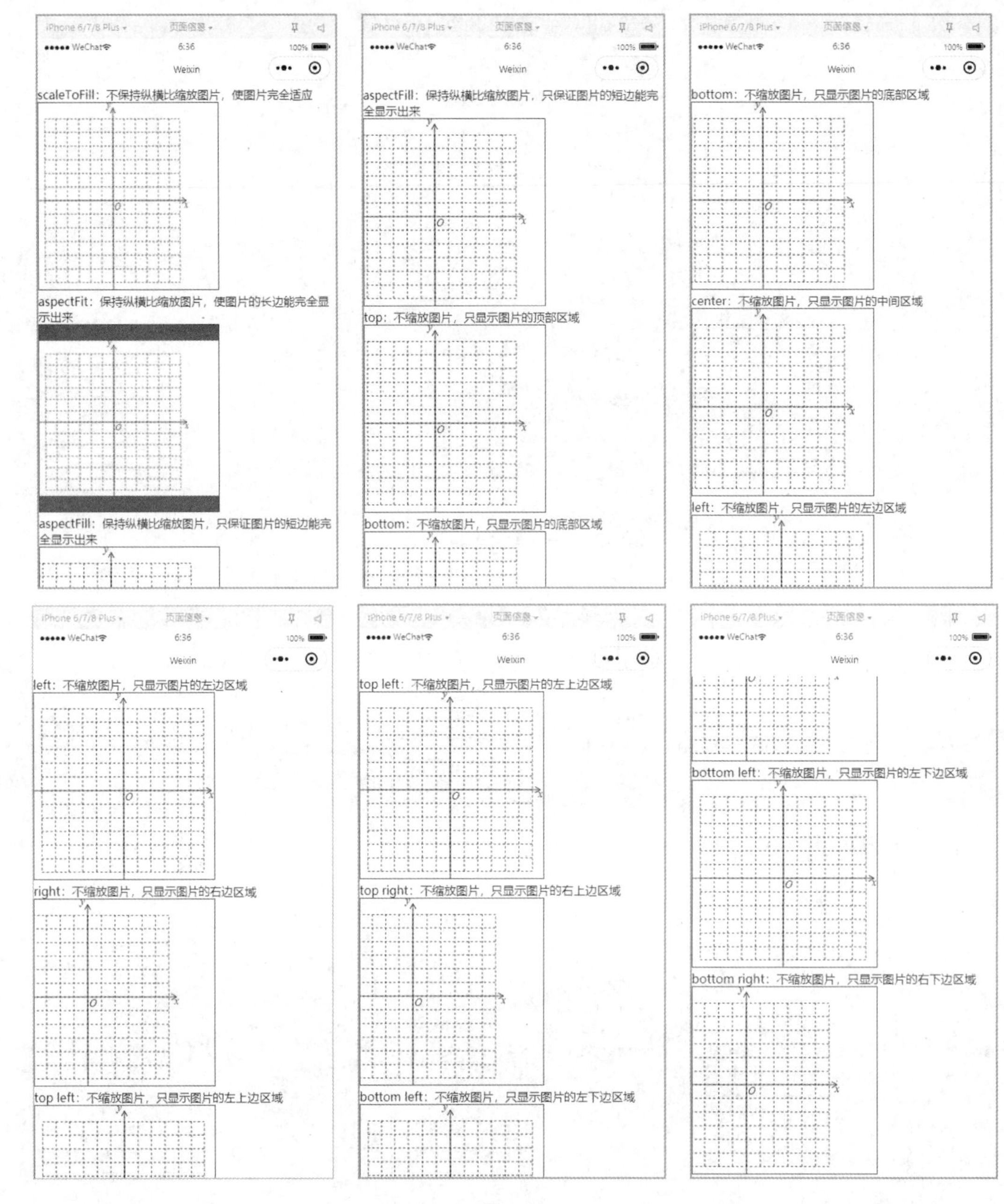

图 4-46

4.6.4　video 组件

video 组件可以实现与视频有关的功能。相关 API：wx.createVideoContext。

video 组件的相关属性如表 4-19 所示。

表 4-19

属　　性	类　　型	默 认 值	必　　填	说　　明	最 低 版 本
src	string		是	要播放视频的资源地址，支持网络路径、本地临时路径、云文件 ID（2.3.0）	1.0.0

续表

属　　性	类　　型	默 认 值	必　　填	说　　明	最 低 版 本
duration	number		否	指定视频时长	1.1.0
controls	boolean	true	否	是否显示默认播放控件（播放/暂停按钮、播放进度、时间）	1.0.0
danmu-list	Array.<object>		否	弹幕列表	1.0.0
danmu-btn	boolean	false	否	是否显示弹幕按钮，只在初始化时有效，不能动态变更	1.0.0
enable-danmu	boolean	false	否	是否展示弹幕，只在初始化时有效，不能动态变更	1.0.0
autoplay	boolean	false	否	是否自动播放	1.0.0
loop	boolean	false	否	是否循环播放	1.4.0
muted	boolean	false	否	是否静音播放	1.4.0
initial-time	number	0	否	指定视频初始播放位置	1.6.0
page-gesture	boolean	false	否	在非全屏模式下，是否开启亮度与音量调节手势（现已废弃，见 vslide-gesture）	1.6.0
direction	number		否	设置全屏时视频的方向，不指定则根据宽高比自动判断。 0：正常竖向； 90：屏幕逆时针 90 度； -90：屏幕顺时针 90 度	1.7.0
show-progress	boolean	true	否	若不设置，则宽度大于 240 时才会显示	1.9.0
show-fullscreen-btn	boolean	true	否	是否显示全屏按钮	1.9.0
show-play-btn	boolean	true	否	是否显示视频底部控制栏的播放按钮	1.9.0
show-center-play-btn	boolean	true	否	是否显示视频中间的播放按钮	1.9.0
enable-progress-gesture	boolean	true	否	是否开启控制进度的手势	1.9.0
object-fit	string	contain	否	当视频大小与 video 容器大小不一致时，视频的表现形式。取值包括 contain（包含）、fill（填充）和 cover（覆盖）	1.0.0
poster	string		否	视频封面的图片网络资源地址或云文件 ID（2.3.0）。若 controls 属性值为 false ，则设置 poster 无效	1.0.0
show-mute-btn	boolean	false	否	是否显示静音按钮	2.4.0
title	string		否	视频的标题，全屏时在顶部展示	2.4.0
play-btn-position	string	bottom	否	播放按钮的位置。 bottom：controls bar 上； center：视频中间	2.4.0
enable-play-gesture	boolean	false	否	是否开启播放手势，即双击切换播放/暂停	2.4.0
auto-pause-if-navigate	boolean	true	否	当跳转到本小程序的其他页面时，是否自动暂停本页面的视频播放	2.5.0
auto-pause-if-open-native	boolean	true	否	当跳转到其他微信原生页面时，是否自动暂停本页面的视频播放	2.5.0
vslide-gesture	boolean	false	否	在非全屏模式下，是否开启亮度与音量调节手势（同 page-gesture）	2.6.2
vslide-gesture-in-fullscreen	boolean	true	否	在全屏模式下，是否开启亮度与音量调节手势	2.6.2

续表

属　　性	类　　型	默 认 值	必　　填	说　　明	最 低 版 本
ad-unit-id	string		是	视频前贴广告单元 ID	2.8.1
poster-for-crawler	string		是	用于给搜索等场景作为视频封面展示，建议使用无播放 icon 的视频封面图，只支持网络地址	
show-casting-button	boolean	false	否	显示投屏按钮。Android 在同层渲染下生效，支持 DLNA 协议；iOS 支持 AirPlay 和 DLNA 协议	2.10.2
picture-in-picture-mode	string/Array		否	设置小窗模式。 []：取消小窗； push：路由 push 时触发小窗； pop：路由 pop 时触发小窗	2.11.0
picture-in-picture-show-progress	boolean	false	否	是否在小窗模式下显示播放进度	2.11.0
enable-auto-rotation	boolean	false	否	开启手机横屏时是否自动全屏，当系统设置开启自动旋转时生效	2.11.0
show-screen-lock-button	boolean	false	否	是否显示锁屏按钮，仅在全屏时显示，锁屏后控制栏的操作	2.11.0
show-snapshot-button	boolean	false	否	是否显示截屏按钮，仅在全屏时显示	2.13.0
show-background-playback-button	boolean	false	否	是否展示后台音频播放按钮	2.14.3
background-poster	string		否	进入后台音频播放后的通知栏图标（Android 独有）	2.14.3
bindplay	eventhandle		否	当开始/继续播放时触发 play 事件	1.0.0
bindpause	eventhandle		否	当暂停播放时触发 pause 事件	1.0.0
bindended	eventhandle		否	当播放到末尾时触发 ended 事件	1.0.0
bindtimeupdate	eventhandle		否	播放进度变化时触发，event.detail = {currentTime, duration} 。触发频率为 250 ms 一次	1.0.0
bindfullscreenchange	eventhandle		否	视频进入和退出全屏时触发，event.detail = {fullScreen, direction}，direction 有效值为 vertical 或 horizontal	1.4.0
bindwaiting	eventhandle		否	视频出现缓冲时触发	1.7.0
binderror	eventhandle		否	视频播放出错时触发	1.7.0
bindprogress	eventhandle		否	加载进度变化时触发，只支持一段加载。event.detail = {buffered}，数据是百分比	2.4.0
bindloadedmetadata	eventhandle		否	视频元数据加载完成时触发。event.detail = {width, height, duration}	2.7.0
bindcontrolstoggle	eventhandle		否	切换 controls 显示隐藏时触发。event.detail = {show}	2.9.5
bindenterpictureinpicture	eventhandler		否	播放器进入小窗	2.11.0
bindleavepictureinpicture	eventhandler		否	播放器退出小窗	2.11.0
bindseekcomplete	eventhandler		否	seek 完成时触发(position iOS 单位为 s，Android 单位为 ms）	2.12.0

（1）JS 程序代码如下。

```
function getRandomColor() {
  const rgb = []
  for (let i = 0; i < 3; ++i) {
    let color = Math.floor(Math.random() * 256).toString(16)
    color = color.length === 1 ? '0' + color : color
    rgb.push(color)
  }
  return '#' + rgb.join('')
}

Page({
  onShareAppMessage() {
    return {
      title: 'video',
      path: 'page/component/pages/video/video'
    }
  },

  onReady() {
    this.videoContext = wx.createVideoContext('myVideo')
  },

  onHide() {

  },

  inputValue: '',
  data: {
    src: '',
    danmuList:
    [{
      text: '第 1s 出现的弹幕',
      color: '#ff0000',
      time: 1
    }, {
      text: '第 3s 出现的弹幕',
      color: '#ff00ff',
      time: 3
    }],
  },

  bindInputBlur(e) {
    this.inputValue = e.detail.value
  },

  bindButtonTap() {
    const that = this
    wx.chooseVideo({
      sourceType: ['album', 'camera'],
      maxDuration: 60,
      camera: ['front', 'back'],
      success(res) {
        that.setData({
          src: res.tempFilePath
```

```
        })
      }
    })
  },

  bindVideoEnterPictureInPicture() {
    console.log('进入小窗模式')
  },

  bindVideoLeavePictureInPicture() {
    console.log('退出小窗模式')
  },

  bindPlayVideo() {
    console.log('1')
    this.videoContext.play()
  },
  bindSendDanmu() {
    this.videoContext.sendDanmu({
      text: this.inputValue,
      color: getRandomColor()
    })
  },

  videoErrorCallback(e) {
    console.log('视频错误信息:')
    console.log(e.detail.errMsg)
  }
})
```

（2）WXML 程序代码如下。

```
<view class="page-body">
  <view class="page-section tc">
    <video
      id="myVideo"
      src="http://localhost/test.mp4"
      binderror="videoErrorCallback"
      danmu-list="{{danmuList}}"
      enable-danmu ="{{true}}"
      danmu-btn ="{{true}}"
      show-center-play-btn='{{true}}'
      show-play-btn="{{true}}"
      controls="{{true}}"
      picture-in-picture-mode="{{['push', 'pop']}}"
      bindenterpictureinpicture='bindVideoEnterPictureInPicture'
      bindleavepictureinpicture='bindVideoLeavePictureInPicture'
    ></video>
    <view style="margin: 30rpx auto" class="weui-label">弹幕内容</view>
    <input bindblur="bindInputBlur" class="weui-input" type="text" placeholder="在此处输入弹幕内容" />
    <button style="margin: 30rpx auto"  bindtap="bindSendDanmu" class="page-body-button" type="primary" formType="submit">发送弹幕</button>
  </view>
</view>
```

（3）video 组件没有 WXSS 程序。

▶ **注意：**

src 需要网络地址，程序编译后效果如图 4-47 所示。

图 4-47

4.7 地图组件

微信小程序的地图组件可实现很多功能，具体如下。

◆ 支持点聚合，适用于 marker（标记点）过多场景。

◆ 支持彩虹蚯蚓线，常用于路线规划场景。

◆ 覆盖物支持调整与其他地图元素的压盖关系。

◆ 支持 marker（如小车）平移动画，适用于轨迹回放场景。

地图组件的相关属性如表 4-20 所示。

表 4-20

属 性	类 型	默 认 值	必 填	说 明	最 低 版 本
longitude	number		是	中心经度	1.0.0
latitude	number		是	中心纬度	1.0.0
scale	number	16	否	缩放级别，取值范围为 3～20	1.0.0
min-scale	number	3	否	最小缩放级别	2.13.0
max-scale	number	20	否	最大缩放级别	2.13.0
markers	Array.<marker>		否	标记点	1.0.0
covers	Array.<cover>		否	即将移除，请使用 markers	1.0.0
polyline	Array.<polyline>		否	路线	1.0.0
circles	Array.<circle>		否	圆	1.0.0

续表

属 性	类 型	默认值	必 填	说 明	最低版本
controls	Array.<control>		否	控件（即将废弃，建议使用 cover-view 代替）	1.0.0
include-points	Array.<point>		否	缩放视野以包含所有给定的坐标点	1.0.0
show-location	boolean	false	否	显示带有方向的当前定位点	1.0.0
polygons	Array.<polygon>		否	多边形	2.3.0
subkey	string		否	个性化地图使用的 key	2.3.0
layer-style	number	1	否	个性化地图配置的 style，不支持动态修改	
rotate	number	0	否	旋转角度，范围为 0～360，地图正北和设备 y 轴角度的夹角	2.5.0
skew	number	0	否	倾斜角度，范围为 0～40，关于 z 轴的倾角	2.5.0
enable-3D	boolean	false	否	展示 3D 楼块（工具暂不支持）	2.3.0
show-compass	boolean	false	否	显示指南针	2.3.0
show-scale	boolean	false	否	显示比例尺（工具暂不支持）	2.8.0
enable-overlooking	boolean	false	否	开启俯视	2.3.0
enable-zoom	boolean	true	否	是否支持缩放	2.3.0
enable-scroll	boolean	true	否	是否支持拖动	2.3.0
enable-rotate	boolean	false	否	是否支持旋转	2.3.0
enable-satellite	boolean	false	否	是否开启卫星地图	2.7.0
enable-traffic	boolean	false	否	是否开启实时路况	2.7.0
enable-poi	boolean	true	否	是否展示 POI（信息点）	2.14.0
enable-building	boolean		否	是否展示建筑物	2.14.0
setting	object		否	配置项	2.8.2
bindtap	eventhandle		否	单击地图时触发，从 2.9.0 版本开始返回经纬度信息	1.0.0
bindmarkertap	eventhandle		否	单击标记点时触发，e.detail = {markerId}	1.0.0
bindlabeltap	eventhandle		否	单击 label 时触发，e.detail = {markerId}	2.9.0
bindcontroltap	eventhandle		否	单击控件时触发，e.detail = {controlId}	1.0.0
bindcallouttap	eventhandle		否	单击标记点对应的气泡时触发，e.detail = {markerId}	1.2.0
bindupdated	eventhandle		否	在地图渲染更新完成时触发	1.6.0
bindregionchange	eventhandle		否	视野发生变化时触发	2.3.0
bindpoitap	eventhandle		否	单击地图 POI 点时触发，e.detail = {name, longitude, latitude}	2.3.0
bindanchorpointtap	eventhandle		否	单击定位标时触发，e.detail = {longitude, latitude}	2.13.0

（1）JS 程序代码如下。

```
Page({
  onReady: function (e) {
    this.mapCtx = wx.createMapContext('myMap')
  },
```

```
  getCenterLocation: function () {
    this.mapCtx.getCenterLocation({
      success: function(res){
        console.log(res.longitude)
        console.log(res.latitude)
      }
    })
  },
  moveToLocation: function () {
    this.mapCtx.moveToLocation()
  },
  translateMarker: function() {
    this.mapCtx.translateMarker({
      markerId: 0,
      autoRotate: true,
      duration: 1000,
      destination: {
        latitude:23.10229,
        longitude:113.3345211,
      },
      animationEnd() {
        console.log('animation end')
      }
    })
  },
  includePoints: function() {
    this.mapCtx.includePoints({
      padding: [10],
      points: [{
        latitude:23.10229,
        longitude:113.3345211,
      }, {
        latitude:23.00229,
        longitude:113.3345211,
      }]
    })
  }
})
```

（2）WXML 程序代码如下。

```
<map id="myMap" show-location />
<button type="primary" size="mini" bindtap="getCenterLocation">获取位置</button>
<button type="primary" size="mini" bindtap="moveToLocation">移动位置</button>
<button type="primary" size="mini" bindtap="translateMarker">移动标注</button>
<button type="primary" size="mini" bindtap="includePoints">缩放视野展示所有经纬度
</button>
```

（3）WXSS 程序代码如下。

```
button{
  margin: 10rpx;
}
```

程序编译后效果如图 4-48、图 4-49 所示。

iPhone 5 100% ...
页面信息
WeChat
7:22
100%
Weixin
京广中心
朝阳门外大街
泰康
北街
德全奢品酒店
腾讯地图
获取位置
移动位置
移动标注
缩放视野展示所有经纬度

图 4-48

iPhone 5 100% ...
页面信息
WeChat
7:22
100%
Weixin
海珠区
长洲岛
大石
S111
腾讯地图
获取位置
移动位置
移动标注
缩放视野展示所有经纬度

图 4-49

第 5 章　小程序自定义组件

本章重点讲解微信小程序中的自定义组件。从小程序基础库版本 1.6.3 开始，小程序支持简洁的组件化编程。所有自定义组件相关特性都需要在基础库 1.6.3 或更高版本上实现。

读者可将页面内的功能模块抽象成自定义组件，以便在不同的页面中重复使用；也可以将复杂的页面拆分成多个低耦合的模块，以方便代码维护。

自定义组件在使用时与基础组件非常相似。

5.1　创建自定义组件

与页面相似，自定义组件也包含 JSON、WXML、WXSS、JS 共 4 个文件。单击鼠标右键，在弹出的快捷菜单中选择“新建 Component”命令，即可快速地创建一个自定义组件，如图 5-1 所示。

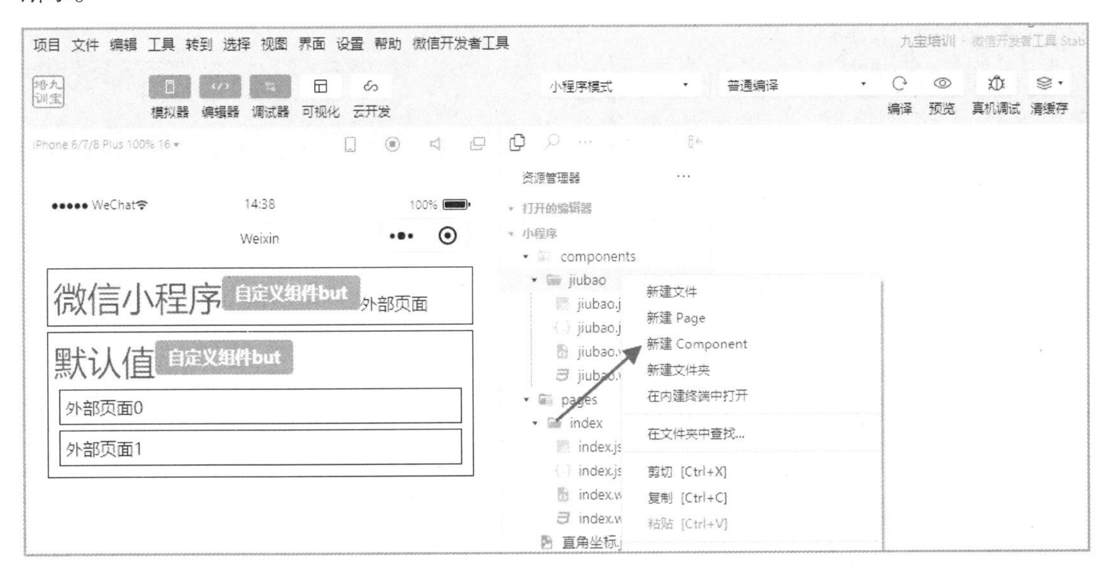

图 5-1

注意：

（1）因为 WXML 节点标签名只能是小写字母、中画线和下画线的组合，所以自定义组件的标签名也只能包含这些字符。

（2）自定义组件同样可以引用自定义组件，引用方法类似于页面对自定义组件的引用，使用 usingComponents 字段。

（3）自定义组件和页面所在的项目根目录名称不能以“wx-”为前缀，否则会报错。

1. JSON 文件

在 JSON 文件中需要进行自定义组件声明。将 component 字段设为 true， 即可将这一组件设置为自定义组件，代码如下。

```
{
  "component": true
}
```

2. JS 文件

与小程序页面不同的是，小程序自定义组件的 JS 文件中需要定义 Component 函数。Component 构造器可用于定义组件，调用 Component 构造器时可以指定组件的属性、数据、方法等。参考代码结构如下。

```
Component({
  /**
   * 组件的属性列表
   */
  properties: {
    arg: {
      type: String,
      value: '默认值',
    }
  },

  /**
   * 组件的初始数据
   */
  data: {

  },

  /**
   * 组件的方法列表
   */
  methods: {
    click(){
      console.log('这是自定义组件',this.properties.arg);
    }
  }
})
```

使用 Component(Object object) 创建自定义组件时，会接收一个 Object 类型的参数。该参数的描述如表 5-1 所示。

表 5-1

定义段	类型	必填	描述	最低版本
properties	Object Map	否	组件的对外属性，是属性名到属性设置的映射表	
data	Object	否	组件的内部数据，和 properties 一同用于组件的模板渲染	
observers	Object	否	组件数据字段监听器，用于监听 properties 和 data 的变化	2.6.1

续表

定义段	类型	必填	描述	最低版本
Methods	Object	否	组件的方法，包括事件响应函数和任意的自定义方法	
behaviors	String Array	否	类似于 mixins 和 traits 的组件间代码复用机制	
created	Function	否	组件生命周期函数，在组件实例刚刚被创建时执行，注意此时不能调用 setData	
attached	Function	否	组件生命周期函数，在组件实例进入页面节点树时执行	
ready	Function	否	组件生命周期函数，在组件布局完成后执行	
moved	Function	否	组件生命周期函数，在组件实例被移动到节点树另一个位置时执行	
detached	Function	否	组件生命周期函数，在组件实例被从页面节点树移除时执行	
relations	Object	否	组件间关系定义	
externalClasses	String Array	否	组件接受的外部样式类	
options	Object Map	否	一些选项（文档中介绍相关特性时会涉及具体的选项设置，这里暂不列举）	
lifetimes	Object	否	组件生命周期声明对象	2.2.3
pageLifetimes	Object	否	组件所在页面的生命周期声明对象	2.2.3
definitionFilter	Function	否	定义段过滤器，用于自定义组件扩展	2.2.3

生成的组件实例可以在组件的方法、生命周期函数和属性 observer 中通过 this 访问。自定义组件中通常包含一些通用属性和方法，如表 5-2、表 5-3 所示。

表 5-2

属性名	类型	描述
is	String	组件的文件路径
id	String	节点 id
dataset	String	节点 dataset
data	Object	组件数据，包括内部数据和属性值
properties	Object	组件数据，包括内部数据和属性值（与 data 一致）

表 5-3

方法名	参数	描述	最低版本
setData	Object newData	设置 data 并执行视图层渲染	
hasBehavior	Object behavior	检查组件是否具有 behavior（检查时会递归检查被直接或间接引入的所有 behavior）	
triggerEvent	String name, Object detail, Object options	触发事件	
createSelectorQuery		创建一个 SelectorQuery 对象，选择器选取范围为这个组件实例内	
createIntersectionObserver		创建一个 IntersectionObserver 对象，选择器选取范围为这个组件实例内	
createMediaQueryObserver		创建一个 MediaQueryObserver 对象	2.11.1

续表

方法名	参数	描述	最低版本
selectComponent	String selector	使用选择器选择组件实例节点，返回匹配到的第一个组件实例对象（会被 wx://component-export 影响）	
selectAllComponents	String selector	使用选择器选择组件实例节点，返回匹配到的全部组件实例对象组成的数组（会被 wx://component-export 影响）	
selectOwnerComponent		选取当前组件节点所在的组件实例（即组件的引用者），返回它的组件实例对象（会被 wx://component-export 影响）	2.8.2
getRelationNodes	String relationKey	获取这个关系所对应的所有关联节点	
groupSetData	Function callback	立刻执行 callback，其中的多个 setData 之间不会触发界面绘制（只有某些特殊场景中需要，如用于在不同组件同时 setData 时进行界面绘制同步）	2.4.0
getTabBar		返回当前页面的 custom-tab-bar 的组件实例	2.6.2
getPageId		返回页面标识符（一个字符串），可以用来判断几个自定义组件实例是否在同一个页面内	2.7.1
animate	String selector, Array keyframes, Number duration, Function callback	执行关键帧动画	2.9.0
clearAnimation	String selector, Object options, Function callback	清除关键帧动画	2.9.0
setUpdatePerformanceListener	Object options, Function listener	更新关键帧动画	2.12.0

properties 属性用来定义组件中的相关变量，如表 5-4 所示。

表 5-4

定义段	类型	必填	描述	最低版本
type		是	属性的类型	
optionalTypes	Array	否	属性的类型（可以指定多个）	2.6.5
value		否	属性的初始值	
observer	Function	否	属性值变化时的回调函数	

注意：

（1）使用 this.data 可以获取内部数据和属性值；但直接修改它不会将变更应用到界面上，应使用 setData 修改。

（2）生命周期函数无法在组件方法中通过 this 访问。

（3）属性名应避免以 data 开头，即不要命名成 dataXyz 的形式，因为在 WXML 中，data-xyz="" 会被作为节点 dataset 来处理，而不是组件属性。

（4）在定义和使用一个组件时，属性名和 data 字段相互不能冲突（尽管它们位于不同的定义段中）。

（5）从基础库 2.0.9 开始，对象类型的属性和 data 字段中都可以包含函数类型的子字段，即可以通过对象类型的属性字段来传递函数。低于这一版本的基础库不支持这一特性。

3．WXML 文件

组件模板的写法与页面模板相同。组件模板与组件数据结合后生成的节点树将被插入组件的引用位置上。

在组件模板中可以提供一个<slot>节点，用于承载组件引用时提供的子节点。示例程序如下。

```
<view>
  <text class="component">{{arg}}</text>
  <button size="mini" type="primary" bindtap="click">自定义组件 but</button>
  <slot name='slot0'></slot>
  <slot name='slot1'></slot>
  <slot></slot>
</view>
```

4．WXSS 文件

组件对应 WXSS 文件的样式，只对组件 WXML 内的节点生效。编写组件样式时，需要注意以下几点。

（1）组件和引用组件的页面中不能使用 id 选择器（#a）、属性选择器（[a]）和标签名选择器，需改用 class 选择器。

（2）应尽量避免在组件和引用组件的页面中使用后代选择器（.a .b），因为在极端情况下会出现一些非预期的表现。

（3）子元素选择器（.a>.b）只能用于 view 组件与其子节点之间，用于其他组件则可能导致非预期的情况。

（4）继承样式会从组件外继承到组件内，如 font、color。

除继承样式之外， app.wxss 中的样式、组件所在页面的样式对自定义组件无效（除非更改组件样式隔离选项）。

```
#a { }              /* 在组件中不能使用 */
{ }                 /* 在组件中不能使用 */
button { }          /* 在组件中不能使用 */
.a > .b { }         /* 除非 .a 是 view 组件节点，否则不一定会生效 */
```

示例程序如下。

```
.component{
  color: red;
  font-size: xx-large;
}
view{
  margin: 10rpx;
  padding: 10rpx;
  border: 1px black solid;
}
```

5.2 引用页面

使用已注册的自定义组件前，要在页面的 JSON 文件中进行引用声明。此时需要提供每个自定义组件的标签名和对应的自定义组件的文件路径，示例程序如下。

```
"usingComponents": {
  "component-jiubao": "./components/jiubao/jiubao"
}
```

在页面的 WXML 中就可以像使用基础组件一样使用自定义组件。节点名即自定义组件的标签名，节点属性即传递给组件的属性值。

注意：

开发者工具 1.02.1810190 及以上版本支持在 app.json 中声明 usingComponents 字段，此处声明的自定义组件会被视为全局自定义组件，在小程序页面或自定义组件中可以直接使用，无须再做声明。

修改 app.json，程序代码如下。

```
{
  "pages": [
    "pages/index/index"
  ],
  "window": {
    "backgroundTextStyle": "light",
    "navigationBarBackgroundColor": "#fff",
    "navigationBarTitleText": "Weixin",
    "navigationBarTextStyle": "black"
  },
  "style": "v2",
  "sitemapLocation": "sitemap.json",
  "usingComponents": {
    "component-jiubao": "./components/jiubao/jiubao"
  }
}
```

仔细观察上述代码，会发现其中增加了以下代码：

```
"usingComponents": {
  "component-jiubao": "./components/jiubao/jiubao"
}
```

亦可以针对 index.json 做以下修改。

```
{
  "usingComponents": {
    "component-jiubao": "../../components/jiubao/jiubao"
  }
}
```

此时的文件结构如图 5-2 所示。

图 5-2

下面给出引用页面的程序清单。

（1）JSON 文件代码如下。

```
{
  "usingComponents": {
    "component-jiubao": "../../components/jiubao/jiubao"
  }
}
```

（2）JS 文件代码如下。

```
Page({

  /**
   * 页面的初始数据
   */
  data: {

  },

  /**
   * 生命周期函数--监听页面加载
   */
  onLoad: function (options) {

  },

  /**
   * 生命周期函数--监听页面初次渲染完成
   */
  onReady: function () {

  },
```

```
  /**
   * 生命周期函数--监听页面显示
   */
  onShow: function () {

  },

  /**
   * 生命周期函数--监听页面隐藏
   */
  onHide: function () {

  },

  /**
   * 生命周期函数--监听页面卸载
   */
  onUnload: function () {

  },

  /**
   * 页面相关事件处理函数--监听用户下拉动作
   */
  onPullDownRefresh: function () {

  },

  /**
   * 页面上拉触底事件的处理函数
   */
  onReachBottom: function () {

  },

  /**
   * 用户单击右上角分享
   */
  onShareAppMessage: function () {

  }
})
```

（3）WXML 文件代码如下。

```
<component-jiubao arg="微信小程序">外部页面</component-jiubao>
<component-jiubao>
  <view slot='slot0'>外部页面 0</view>
  <view slot='slot1'>外部页面 1</view>
</component-jiubao>
```

（4）引用页面没有 WXSS 文件代码。

5.3　程序解读

下面对自定义组件相关示例程序进行解读。

5.3.1　引用自定义组件

前面的示例程序编译后的效果如图 5-3 所示。

图 5-3

index.json 文件中给出了微信小程序自定义组件的引用方式。自定义组件名称是 component-jiubao，引用路径是../../components/jiubao/jiubao。相关代码如下。

```
{
  "usingComponents": {
    "component-jiubao": "../../components/jiubao/jiubao"
  }
}
```

此时，index 页面可以使用名为 component-jiubao，指向../../components/jiubao/jiubao 的自定义组件。

index.wxml 使用名为 component-jiubao 的自定义组件，相关程序如下。

```
<component-jiubao arg="微信小程序">外部页面</component-jiubao>
<component-jiubao>
  <view slot='slot0'>外部页面 0</view>
  <view slot='slot1'>外部页面 1</view>
</component-jiubao>
```

5.3.2　slot

上述程序给出了两种应用 slot 的方式。一种是单 slot 节点形式，代码如下。

```
<component-jiubao arg="微信小程序">外部页面</component-jiubao>
```

另一种是多 slot 节点形式，代码如下。

```
<component-jiubao> <view slot='slot0'>外部页面 0</view><view slot='slot1'>外部页面
1</view></component-jiubao>
```

对于单 slot 节点形式，自定义组件将在 slot 标签中直接显示引用页面，相关程序代码如下。

```
<view>
  <text class="component">{{arg}}</text>
  <button size="mini" type="primary" bindtap="click">自定义组件 but</button>
  <slot name='slot0'></slot>
  <slot name='slot1'></slot>
  <slot></slot>
</view>
```

对于多 slot 节点形式，需要在 JS 文件中定义 multipleSlots: true，相关程序代码如下。

```
Component({
  /**
   * 组件的属性列表
   */
  properties: {
    arg: {
      type: String,
      value: '默认值',
    },
  },

  options: {
    multipleSlots: true // 在组件定义选项中启用多 slot 支持
  },

  /**
   * 组件的初始数据
   */
  data: {

  },

  /**
   * 组件的方法列表
   */
  methods: {
    click(){
      console.log('这是自定义组件',this.properties.arg);
    }
  }
})
```

自定义组件 WXML 的 slot 节点，将按照 name 对应显示信息，相关程序代码如下。

```
<view>
  <text class="component">{{arg}}</text>
  <button size="mini" type="primary" bindtap="click">自定义组件 but</button>
  <slot name='slot0'></slot>
  <slot name='slot1'></slot>
  <slot></slot>
</view>
```

5.3.3　自定义组件样式

微信小程序自定义组件的样式定义与普通页面相似。使用“<text class="component">{{arg}}</text>”定义 text 引用的 component 样式，示例程序如下。

```
<view>
  <text class="component">{{arg}}</text>
  <button size="mini" type="primary" bindtap="click">自定义组件 but</button>
  <slot name='slot0'></slot>
  <slot name='slot1'></slot>
  <slot></slot>
</view>
```

对应自定义组件的样式如下。

```
.component{
  color: red;
  font-size: xx-large;
}
view{
  margin: 10rpx;
  padding: 10rpx;
  border: 1px black solid;
}
```

5.3.4　自定义组件事件

微信小程序自定义组件的事件定义与普通页面很相似。例如，通过“<button size= "mini" type="primary" bindtap="click">自定义组件 but</button>”定义 button 的 tap 事件，下面修改 JS 文件实现这个函数。

（1）WXML 文件代码如下。

```
<view>
  <text class="component">{{arg}}</text>
  <button size="mini" type="primary" bindtap="click">自定义组件 but</button>
  <slot name='slot0'></slot>
  <slot name='slot1'></slot>
  <slot></slot>
</view>
```

（2）JS 文件代码如下。

```
Component({
  /**
   * 组件的属性列表
   */
  properties: {
    arg: {
      type: String,
      value: '默认值',
    },
```

```
  },

  options: {
    multipleSlots: true // 在组件定义选项中启用多slot支持
  },

  /**
   * 组件的初始数据
   */
  data: {

  },

  /**
   * 组件的方法列表
   */
  methods: {
    click(){
      console.log('这是自定义组件',this.properties.arg);
    }
  }
})
```

第 6 章　小程序 API

API 是微信小程序学习的另一个重要知识点。本章重点介绍微信小程序中常用的一些 API，其他 API 会在各章节的示例讲解中随用随讲。

6.1　基础 API

下面对微信小程序中两个常用的基础 API 进行讲解。

6.1.1　boolean wx.canIUse(string schema)

boolean wx.canIUse(string schema)的作用是判断小程序的 API、回调、参数、组件等是否可在当前版本中使用。

对于一些较新的功能组件和 API，需要在使用时先判断用户的手机环境（即手机微信小程序版本）是否支持。如果不支持，可以提供降级解决方案，或者给出相关提示。

例如，判断“wx.showToast”是否可用，代码如下。

```
onLoad: function (options) {
  console.log(wx.canIUse('showToast'))
},
```

返回结果如下。

```
true
```

其中，参数 schema 的格式是 ${API}.${method}.${param}.${option} 或 ${component}.${attribute}.${option}。相关说明如下。

- ${API} 表示 API 名字。
- ${method} 表示调用方式，有效值为 return、success、object 和 callback。
- ${param} 表示参数或返回值。
- ${option} 表示参数的可选值或返回值的属性。
- ${component} 表示组件名称。
- ${attribute} 表示组件属性。
- ${option} 表示组件属性的可选值。

6.1.2　Object wx.getSystemInfoSync()

在微信小程序中，部分 API 分同步和异步两个版本。以获取系统信息为例，同步版本是 wx.getSystemInfoSync()，异步版本是 wx.getSystemInfoAsync(Object object)。

注意：

（1）同步指的是当前程序必须执行结束后方能执行后续程序，异步指的是当前程序不必等待最终执行结束即可执行后续程序。

（2）不是所有的微信小程序都同时提供同步 API 与异步 API。

（3）部分 API 在腾讯官方文档中未明确说明是同步还是异步的。

（4）部分 API 以同步、异步、老旧版本的形式提供。

下面以 wx.getSystemInfoSync()为例，介绍同步 API 和异步 API 的调用方式，其他 API 的调用方式可以借鉴。

调用同步 wx.getSystemInfoSync() 的示例程序如下。

```
onLoad: function (options) {
  let str = JSON.stringify(wx.getSystemInfoSync())
  console.log(str)
},
```

调用执行结果如下。


```
{"model":"iPhone 6/7/8
Plus","pixelRatio":3,"windowWidth":414,"windowHeight":672,"system":"iOS
10.0.1","language":"zh_CN","version":"7.0.4","screenWidth":414,"screenHeight":736,"SDK
Version":"2.16.0","brand":"devtools","fontSizeSetting":16,"benchmarkLevel":1,"batteryL
evel":100,"statusBarHeight":20,"safeArea":{"top":20,"left":0,"right":414,"bottom":736,
"width":414,"height":716},"deviceOrientation":"portrait","platform":"devtools","enable
Debug":false,"devicePixelRatio":3}
```


将上述结果进行格式化，可得如下结果。


```
{
"model": "iPhone 6/7/8 Plus",
"pixelRatio": 3,
"windowWidth": 414,
"windowHeight": 672,
"system": "iOS 10.0.1",
"language": "zh_CN",
"version": "7.0.4",
"screenWidth": 414,
"screenHeight": 736,
"SDKVersion": "2.16.0",
"brand": "devtools",
"fontSizeSetting": 16,
"benchmarkLevel": 1,
"batteryLevel": 100,
"statusBarHeight": 20,
"safeArea": {
    "top": 20,
    "left": 0,
    "right": 414,
    "bottom": 736,
    "width": 414,
    "height": 716
},
"deviceOrientation": "portrait",
"platform": "devtools",
"enableDebug": false,
"devicePixelRatio": 3
}
```


调用异步 wx.getSystemInfoSync() 的示例程序如下。

```
onLoad: function (options) {
```

```
    wx.getSystemInfoAsync({
      success (res){
        let str = JSON.stringify(wx.getSystemInfoSync())
        console.log(str)
      }
    })
  },
```

调用执行结果如下。

```
  {"model":"iPhone 6/7/8
Plus","pixelRatio":3,"windowWidth":414,"windowHeight":672,"system":"iOS
10.0.1","language":"zh_CN","version":"7.0.4","screenWidth":414,"screenHeight":736,"SDK
Version":"2.16.0","brand":"devtools","fontSizeSetting":16,"benchmarkLevel":1,"batteryL
evel":100,"statusBarHeight":20,"safeArea":{"top":20,"left":0,"right":414,"bottom":736,
"width":414,"height":716},"deviceOrientation":"portrait","platform":"devtools","enable
Debug":false,"devicePixelRatio":3}
```

将上述结果进行格式化，可得如下结果。

```
  {
  "model": "iPhone 6/7/8 Plus",
  "pixelRatio": 3,
  "windowWidth": 414,
  "windowHeight": 672,
  "system": "iOS 10.0.1",
  "language": "zh_CN",
  "version": "7.0.4",
  "screenWidth": 414,
  "screenHeight": 736,
  "SDKVersion": "2.16.0",
  "brand": "devtools",
  "fontSizeSetting": 16,
  "benchmarkLevel": 1,
  "batteryLevel": 100,
  "statusBarHeight": 20,
  "safeArea": {
      "top": 20,
      "left": 0,
      "right": 414,
      "bottom": 736,
      "width": 414,
      "height": 716
  },
  "deviceOrientation": "portrait",
  "platform": "devtools",
  "enableDebug": false,
  "devicePixelRatio": 3
  }
```

6.1.3　更新微信小程序版本

一般来说，用户打开微信小程序时，系统需要检测当前用户使用的小程序版本是否为最新版本。对于已发布过新版本的微信小程序，旧版小程序应在检测后给出更新提示信息。

检测新版本的功能通常使用 app.js 中的 updateManager.applyUpdate()函数实现，相关程序代码如下。

```
  App({
    onLaunch() {
      const updateManager = wx.getUpdateManager()
```

```
        updateManager.onCheckForUpdate(function (res) {
          console.log('请求完新版本信息的回调',res.hasUpdate)
        })

        updateManager.onUpdateFailed(function () {
          console.log('新版本下载失败')
        })

        updateManager.onUpdateReady(function () {
          wx.showModal({
            title: '更新提示',
            content: '新版本已经准备好，是否重启应用？',
            success: function (res) {
              if (res.confirm) {
                // 新的版本已经下载，调用 applyUpdate 应用新版本并重启
                updateManager.applyUpdate()
              }
            }
          })
        })
      }
    })
```

注意：

（1）在微信开发者工具中，可通过“编译模式”下的“下次编译模拟更新”开关来调试。

（2）小程序的开发版/体验版没有“版本”概念，所以无法在开发版/体验版上测试版本更新情况。

开发测试时，单击菜单功能区的“编译”按钮，在弹出的下拉菜单中选择“添加编译模式”命令，如图 6-1 所示。

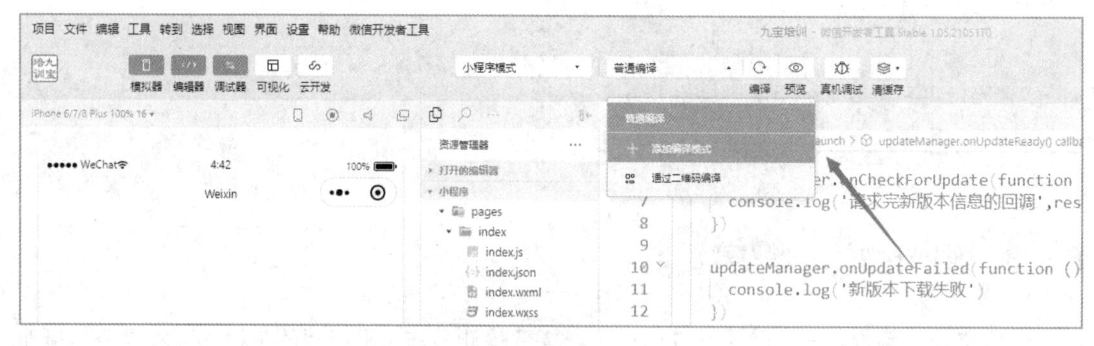

图 6-1

打开“自定义编译条件”对话框，先在“编译设置”处选中“下次编译时模拟更新（需 1.9.90 及以上基础库版本）”复选框，在“更新状态”处选中“成功状态”单选按钮，然后单击“确定”按钮，如图 6-2 所示。

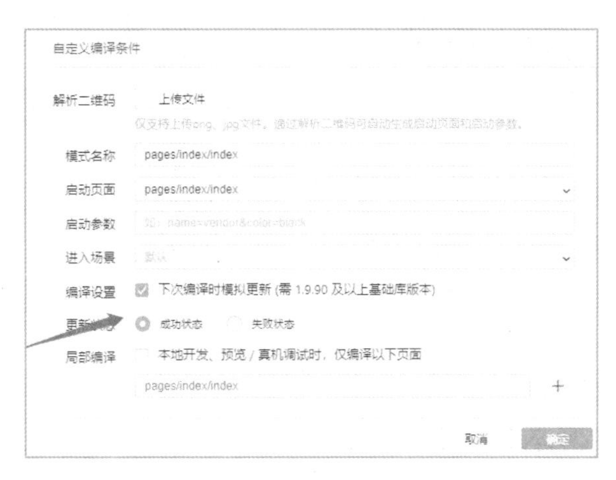

图 6-2

设置完毕后，再次单击“编译”按钮，如图 6-3 所示。

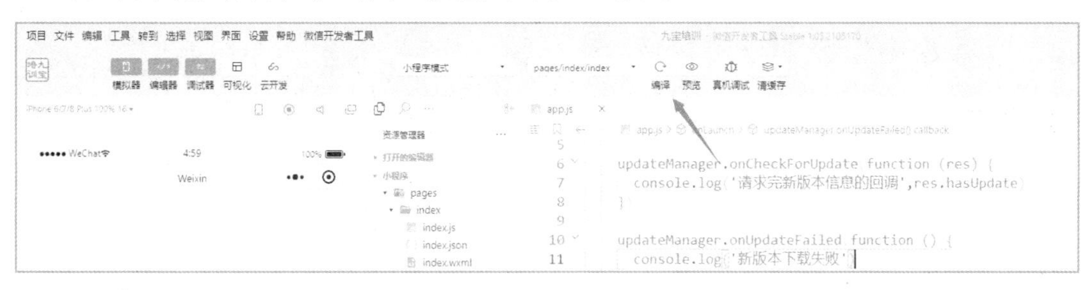

图 6-3

此时，微信小程序模拟器将收到更新提示，如图 6-4 所示。单击“确定”按钮后，将提示重启小程序，如图 6-5 所示。单击“知道了”按钮后小程序进入重启中状态，如图 6-6 所示。

图 6-4

图 6-5

图 6-6

此时，调试器 Console 处会打印出相关信息，如图 6-7 所示。

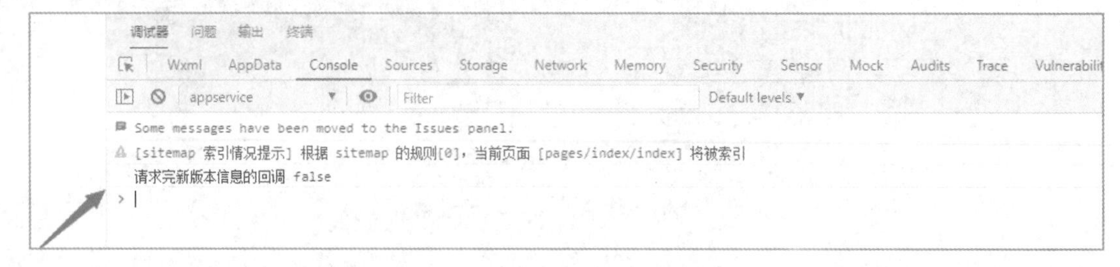

图 6-7

6.1.4 更新微信版本

更新微信版本时需要使用 wx.updateWeChatApp()函数，相关程序代码如下。

```
App({
  onLaunch() {

    const updateManager = wx.getUpdateManager()

    updateManager.onCheckForUpdate(function (res) {
      console.log('请求完新版本信息的回调',res.hasUpdate)
    })

    updateManager.onUpdateFailed(function () {
      console.log('新版本下载失败')
    })

    updateManager.onUpdateReady(function () {
      wx.showModal({
        title: '更新提示',
        content: '新版本已经准备好，是否重启应用？',
        success: function (res) {
          if (res.confirm) {
            // 新的版本已经下载，调用 applyUpdate 应用新版本并重启
            //updateManager.applyUpdate()
            wx.updateWeChatApp({success(){console.log('wx.updateWeChatApp')}})
          }
        }
      })
    })

  }
})
```

注意：

上述程序仅供学习参考。在实际开发中应根据规则选择使用“更新微信小程序”和“更新微信客户端”。

更新微信版本的操作与更新小程序版本相似，同样需要先打开“自定义编译条件”对话框，在其中对“编译设置”和“更新状态”进行设置，然后再次进行编译，此时会弹出如图 6-8 所

示的更新提示信息，单击“确定”按钮后，将给出“当前版本过低，请更新微信”的提示信息，如图 6-9 所示。

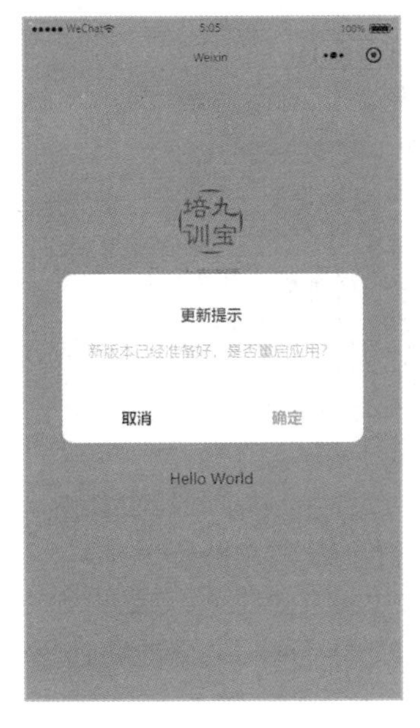

图 6-8

图 6-9

6.2　网络 API

6.2.1　wx.request

wx.request 用于发起 HTTPS 网络请求。最终发送给服务器的数据是 String 类型，如果传入的 data 不是 String 类型，则会被转换成 String 类型。转换规则如下。

- 对于用 GET 方法发送的数据，会将数据转换成 query string（encodeURIComponent(k)=encodeURIComponent(v)&encodeURIComponent(k)=encodeURIComponent(v)...）。
- 对于用 POST 方法发送且 header['content-type'] 为 application/json 的数据，会对数据进行 JSON 序列化。
- 对于用 POST 方法发送且 header['content-type']为 application/x-www-form-urlencoded 的数据，会将数据转换成 query string(encodeURIComponent(k)=encodeURI Component (v)&encodeURIComponent(k)=encodeURIComponent(v)...）。

wx.request 的相关参数如表 6-1 所示。

表 6-1

属　性	类　型	默 认 值	必　填	说　明	最 低 版 本
url	string		是	开发者服务器接口地址	
data	string/object/ArrayBuffer		否	请求的参数	
header	Object		否	设置请求的 header，其中不能设置 Referer。content-type 默认为 application/json	

续表

属　性	类　型	默认值	必　填	说　明	最低版本
timeout	number		否	超时时间，单位为毫秒	2.10.0
method	string	GET	否	HTTP 请求方法。 OPTIONS HTTP：请求 OPTIONS； GET HTTP：请求 GET； HEAD HTTP 请求 HEAD； POST HTTP 请求 POST； PUT HTTP 请求 PUT； DELETE HTTP 请求 DELETE； TRACE HTTP：请求 TRACE； CONNECT HTTP：请求 CONNECT	
dataType	string	json	否	返回的数据格式。 json：返回格式为 JSON，返回后会对返回数据进行 JSON.parse； 其他：不对返回内容进行 JSON.parse	
responseType	string	text	否	响应的数据类型。 text：响应数据为文本； arraybuffer：响应数据为 ArrayBuffer	1.7.0
enableHttp2	boolean	false	否	开启 http2	2.10.4
enableQuic	boolean	false	否	开启 quic	2.10.4
enableCache	boolean	false	否	开启 cache	2.10.4
success	function		否	接口调用成功的回调函数。参见表 6-2	
fail	function		否	接口调用失败的回调函数	
complete	function		否	接口调用结束的回调函数（调用成功、失败都会执行）	

object.success 回调函数的相关参数如表 6-2 所示。

表 6-2

属　性	类　型	说　明	最低版本
data	string/Object/Arraybuffer	开发者服务器返回的数据	
statusCode	number	开发者服务器返回的 HTTP 状态码	
header	Object	开发者服务器返回的 HTTP Response Header	1.2.0
cookies	Array.<string>	开发者服务器返回的 cookies，格式为字符串数组	2.10.0
profile	Object	网络请求过程中的调试信息，参见表 6-3	2.10.4

res.profile 结构如表 6-3 所示。

表 6-3

属　性	类　型	说　明
redirectStart	number	第一个 HTTP 重定向发生时的时间。有跳转且是同域名内部的重定向才有效，否则值为 0

续表

属　性	类　型	说　明
redirectEnd	number	最后一个 HTTP 重定向完成时的时间。有跳转且是同域名内部的重定向才有效，否则值为 0
fetchStart	number	组件准备好使用 HTTP 请求抓取资源的时间，这发生在检查本地缓存之前
domainLookupStart	number	DNS 域名查询开始的时间，如果使用了本地缓存（即无 DNS 查询）或持久连接，则与 fetchStart 值相等
domainLookupEnd	number	DNS 域名查询完成的时间，如果使用了本地缓存（即无 DNS 查询）或持久连接，则与 fetchStart 值相等
connectStart	number	HTTP（TCP）开始建立连接的时间，如果是持久连接，则与 fetchStart 值相等。注意，如果在传输层发生了错误且重新建立连接，则这里显示的是新建立的连接开始的时间
connectEnd	number	HTTP（TCP）完成建立连接的时间（完成握手），如果是持久连接，则与 fetchStart 值相等。注意，如果在传输层发生了错误且重新建立连接，则这里显示的是新建立的连接完成的时间。这里握手结束，包括安全连接建立完成、SOCKS 授权通过
SSLconnectionStart	number	SSL 开始建立连接的时间，如果不是安全连接，则值为 0
SSLconnectionEnd	number	SSL 完成建立连接的时间，如果不是安全连接，则值为 0
requestStart	number	HTTP 请求读取真实文档开始的时间（开始建立连接），包括从本地读取缓存。因连接错误而需要重新连接时，这里显示的也是新建立连接开始的时间
requestEnd	number	HTTP 请求读取真实文档结束的时间
responseStart	number	HTTP 开始接收响应的时间（获取到第一个字节），包括从本地读取缓存
responseEnd	number	HTTP 响应全部接收完成的时间（获取到最后一个字节），包括从本地读取缓存
rtt	number	当次请求连接过程中实时 rtt
estimate_nettype	string	评估的网络状态，取值可能为 slow 2g/2g/3g/4g
httpRttEstimate	number	协议层根据多个请求评估当前网络的 rtt（仅供参考）
transportRttEstimate	number	传输层根据多个请求评估当前网络的 rtt（仅供参考）
downstreamThroughputKbpsEstimate	number	评估当前网络下载的传输速度（kbps）
throughputKbps	number	当前网络的实际下载传输速度（kbps）
peerIP	string	当前请求的 IP
port	number	当前请求的端口
socketReused	boolean	是否复用连接
sendBytesCount	number	发送的字节数
receivedBytedCount	number	收到的字节数

在 wx.request 接口开发状态，建议配置微信开发工具。单击“详情”按钮，在弹出面板的“本地设置”选项卡下选中“不校验合法域名、web-view（业务域名）、TLS 版本以及 HTTPS 证书”复选框，如图 6-10 所示。

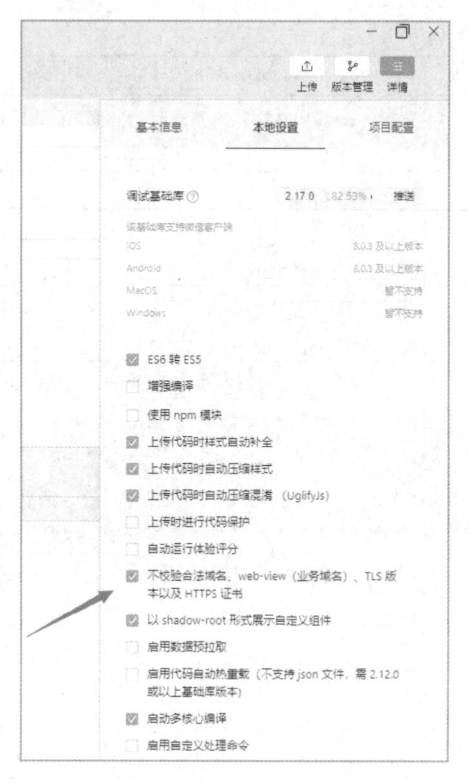

图 6-10

示例程序如下。

```
Page({

  /**
   * 页面的初始数据
   */
  data: {

  },

  /**
   * 生命周期函数--监听页面加载
   */
  onLoad: function (options) {
    wx.request({
      url:'https://www.qq.com/',
      success(res){
        console.log(res.data);
      }})
  },

  /**
   * 生命周期函数--监听页面初次渲染完成
   */
  onReady: function () {

  },

  /**
   * 生命周期函数--监听页面显示
   */
  onShow: function () {

  },
```

```
  /**
   * 生命周期函数--监听页面隐藏
   */
  onHide: function () {

  },

  /**
   * 生命周期函数--监听页面卸载
   */
  onUnload: function () {

  },

  /**
   * 页面相关事件处理函数--监听用户下拉动作
   */
  onPullDownRefresh: function () {

  },

  /**
   * 页面上拉触底事件的处理函数
   */
  onReachBottom: function () {

  },

  /**
   * 用户单击右上角分享
   */
  onShareAppMessage: function () {

  }
})
```

编译微信小程序，观察 Console 执行结果，如图 6-11 所示。

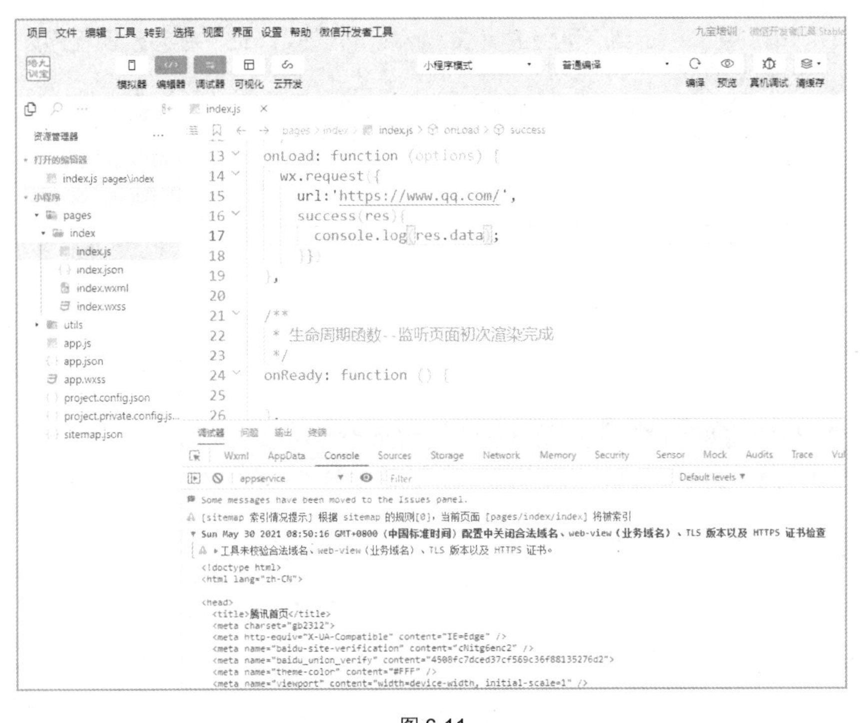

图 6-11

注意：

对于正式开发环境，需要配置正式账号的域名。在微信小程序开发工具中单击“详情”按钮，可在“项目配置”选项卡查看当前可用的域名信息，如图 6-12 所示。

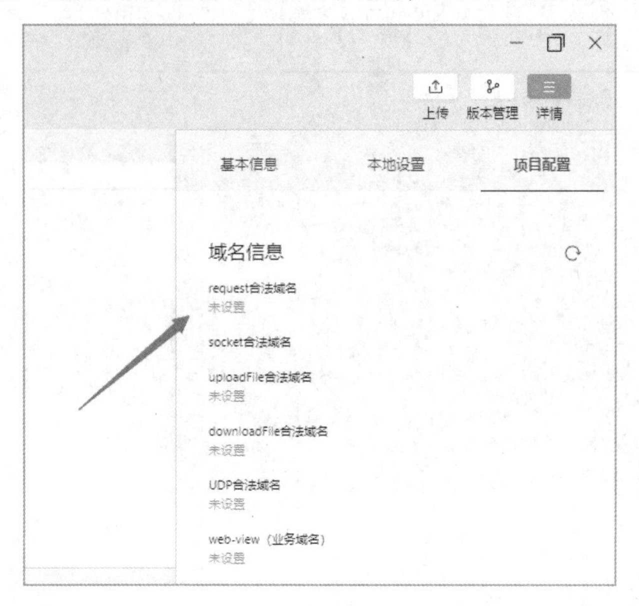

图 6-12

另外，wx.request 是异步请求。异步请求的特点是当请求发起后，无须等待请求结果响应，立即执行后续程序。同步请求的特点是当请求发起后，必须等待请求结果响应。微信小程序中的 API 函数大多采用异步方式。

下面以 wx.request 为例统一进行说明，后续在其他章节中不再赘述。

对于需要将响应结果绑定到变量上的需求，不能直接使用 this。参考示例程序如下。

（1）JS 文件代码如下。

```
Page({

  /**
   * 页面的初始数据
   */
  data: {
    str:''
  },

  /**
   * 生命周期函数--监听页面加载
   */
  onLoad: function (options) {
    wx.request({
      url:'https://www.qq.com/',
      success(res){
        getCurrentPages()[getCurrentPages().length-1].setData({str:res.data})
        console.log(res.data);
      }})
  },
```

```
  /**
   * 生命周期函数--监听页面初次渲染完成
   */
  onReady: function () {

  },

  /**
   * 生命周期函数--监听页面显示
   */
  onShow: function () {

  },

  /**
   * 生命周期函数--监听页面隐藏
   */
  onHide: function () {

  },

  /**
   * 生命周期函数--监听页面卸载
   */
  onUnload: function () {

  },

  /**
   * 页面相关事件处理函数--监听用户下拉动作
   */
  onPullDownRefresh: function () {

  },

  /**
   * 页面上拉触底事件的处理函数
   */
  onReachBottom: function () {

  },

  /**
   * 用户单击右上角分享
   */
  onShareAppMessage: function () {

  }
})
```

（2）WXML 文件代码如下。

```
{{str}}
```

代码解析如下。

（1）JS 文件定义了 str 变量，通过数据绑定，将 str 绑定到 WXML。此外，onLoad 事件定义使用 wx.request 访问 https://www.qq.com/。wx.reuqest 的回调函数使用 getCurrentPages()[getCurrentPages().length-1].setData({str:res.data})将请求 https://www.qq.com/的响应信息绑定到变量 str 上。

（2）getCurrentPages()的作用是获取当前页面栈。数组中，第一个元素为首页，最后一个元素为当前页面。

（3）getCurrentPages().length 的作用是获取当前页面栈数，getCurrentPages()[getCurrentPages(). length-1]获取的是当前页面，也就是 this。

程序最终实现的效果如图 6-13 所示。

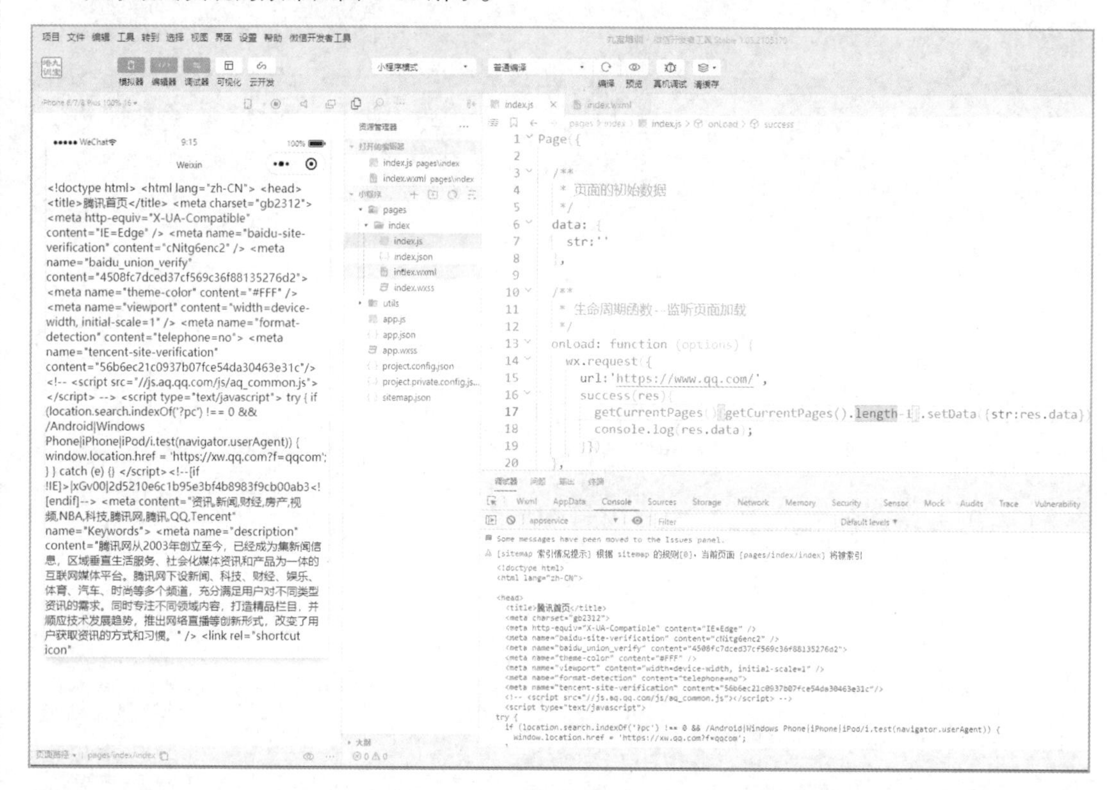

图 6-13

6.2.2 wx.uploadFile

wx.uploadFile 用于将本地资源上传到开发者服务器。当页面通过 wx.chooseImage 等接口获取一个本地资源的临时文件路径后，可通过此接口将本地资源上传到指定服务器。客户端发起一个 HTTPS POST 请求，其中 content-type 为 multipart/form-data。

wx.uploadFile 相关参数如表 6-4 所示。

表 6-4

属　性	类　型	默认值	必　填	说　明	最低版本
url	string		是	开发者服务器地址	
filePath	string		是	要上传文件资源的路径（本地路径）	
name	string		是	文件对应的 key，开发者在服务端可以通过这个 key 获取文件的二进制内容	

续表

属　　性	类　　型	默 认 值	必　　填	说　　明	最 低 版 本
header	Object		否	HTTP 请求 Header，Header 中不能设置 Referer	
formData	Object		否	HTTP 请求中其他额外的 form data	
timeout	number		否	超时时间，单位为毫秒	2.10.0
success	function		否	接口调用成功的回调函数，参见表 6-5	
fail	function		否	接口调用失败的回调函数	
complete	function		否	接口调用结束的回调函数（调用成功、失败都会执行）	

object.success 回调函数的相关参数如表 6-5 所示。

表 6-5

属　　性	类　　型	说　　明
data	string	开发者服务器返回的数据
statusCode	number	开发者服务器返回的 HTTP 状态码

微信小程序示例代码如下。

```
Page({

  /**
   * 页面的初始数据
   */
  data: {

  },

  /**
   * 生命周期函数--监听页面加载
   */
  onLoad: function (options) {
    wx.chooseImage({
      success: function (res) {
        var tempFilePaths = res.tempFilePaths
        console.log(tempFilePaths);
        wx.uploadFile({
          url: 'http://localhost/wx/Upload', //仅为示例，非真实的接口地址
          filePath: tempFilePaths[0],
          name: 'file',
          formData: {
            'arg': '九宝'
          },
          success: function (res) {
            var data = res.data;
            var pages = getCurrentPages();
            var page = pages[pages.length - 1];
            page.setData({ text: data });
          },
          fail: function (res) {
            console.log(res);
          }
        })
      }
    })
  },

  /**
```

```
   * 生命周期函数--监听页面初次渲染完成
   */
  onReady: function () {

  },

  /**
   * 生命周期函数--监听页面显示
   */
  onShow: function () {

  },

  /**
   * 生命周期函数--监听页面隐藏
   */
  onHide: function () {

  },

  /**
   * 生命周期函数--监听页面卸载
   */
  onUnload: function () {

  },

  /**
   * 页面相关事件处理函数--监听用户下拉动作
   */
  onPullDownRefresh: function () {

  },

  /**
   * 页面上拉触底事件的处理函数
   */
  onReachBottom: function () {

  },

  /**
   * 用户单击右上角分享
   */
  onShareAppMessage: function () {

  }
})
```

对应的服务端程序如下（以 Java 语言服务端为例）。

```
package wx;

import java.io.BufferedInputStream;
import java.io.File;
import java.io.FileOutputStream;
import java.io.IOException;
import javax.servlet.ServletException;
import javax.servlet.annotation.MultipartConfig;
import javax.servlet.annotation.WebServlet;
import javax.servlet.http.HttpServlet;
import javax.servlet.http.HttpServletRequest;
```

```
    import javax.servlet.http.HttpServletResponse;

    @WebServlet("/Upload")

    public class WX extends HttpServlet {
    protected void doPost(HttpServletRequest request, HttpServletResponse response)
throws ServletException, IOException {
        request.setCharacterEncoding("UTF-8");
        response.setCharacterEncoding("UTF-8");
        System.out.println("doPost");
        BufferedInputStream bin = new
BufferedInputStream(request.getPart("file").getInputStream());
        FileOutputStream fileout = new FileOutputStream(new File("e:/wx.jpg"));

        byte[] b = new byte[1024];
        int x = -1;
        while(-1!= (x = bin.read(b,0,b.length))){
            fileout.write(b,0,x);
            fileout.flush();
        }
        fileout.close();
        bin.close();
        response.getWriter().print(request.getParameter("arg"));
        System.out.println(request.getParameter("arg"));
    }
    }
```

编译微信小程序后，选择文件上传，被选文件就会上传到服务器上，如图 6-14 所示。

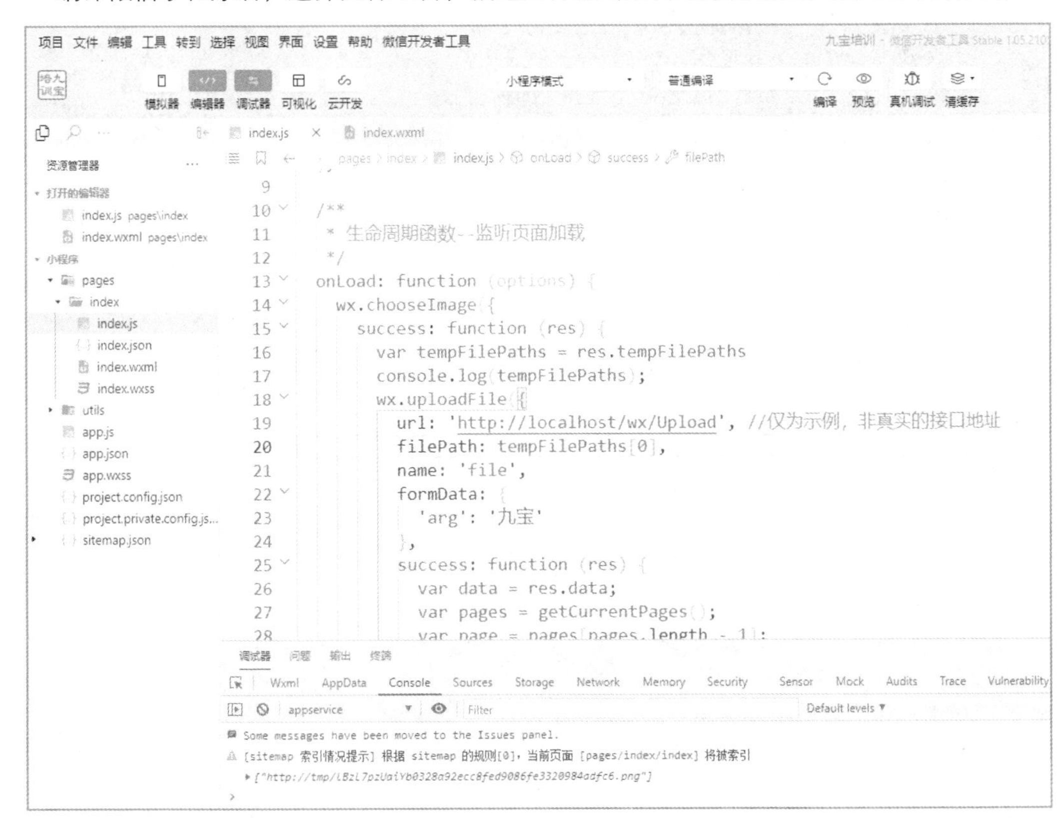

图 6-14

此时，Console 打印的是微信小程序本地文件的路径。

服务器接收文件的同时，接收到参数 arg 的值为“九宝”，如图 6-15 所示。

```
16 public class WX extends HttpServlet {
17     protected void doPost(HttpServletRequest request, HttpServletRespons
18         request.setCharacterEncoding("UTF-8");
19         response.setCharacterEncoding("UTF-8");
20         System.out.println("doPost");
21         BufferedInputStream bin = new BufferedInputStream(request.getPar
22         FileOutputStream fileout = new FileOutputStream(new File("e:/wx.
23         byte[] b = new byte[1024];
24         int x = -1;
25         while(-1!= (x = bin.read(b,0,b.length))){
26             fileout.write(b,0,x);
27             fileout.flush();
28         }
29         fileout.close();
30         bin.close();
31         response.getWriter().print(request.getParameter("arg"));
32         System.out.println(request.getParameter("arg"));
33     }
34 }
35
```

```
信息: 初始化协议处理器 ["http-nio-80"]
5月 30, 2021 10:22:19 上午 org.apache.catalina.startup.Catalina load
信息: 服务器在[628]毫秒内初始化
5月 30, 2021 10:22:19 上午 org.apache.catalina.core.StandardService startInternal
信息: 正在启动服务[Catalina]
5月 30, 2021 10:22:19 上午 org.apache.catalina.core.StandardEngine startInternal
信息: 正在启动 Servlet 引擎：[Apache Tomcat/9.0.39]
5月 30, 2021 10:22:20 上午 org.apache.catalina.util.SessionIdGeneratorBase createSecureRandom
警告: 使用[SHA1PRNG]创建会话ID生成的SecureRandom实例花费了[169]毫秒。
5月 30, 2021 10:22:20 上午 org.apache.coyote.AbstractProtocol start
信息: 开始协议处理句柄["http-nio-80"]
5月 30, 2021 10:22:20 上午 org.apache.catalina.startup.Catalina start
信息: [430]毫秒后服务器启动
doPost
九宝
```

图 6-15

6.2.3 wx.downloadFile

wx.downloadFile 用于实现下载文件资源到本地。在客户端直接发起一个 HTTP GET 请求，将返回文件的本地临时路径。

下面通过示例程序来说明 wx.downloadFile 的使用方法。

```
Page({

  /**
   * 页面的初始数据
   */
  data: {

  },

  /**
   * 生命周期函数--监听页面加载
   */
  onLoad: function (options) {
    wx.downloadFile({
      url: 'http://localhost/wx/wx.jpg', //仅为示例，并非真实的资源
      success: function (res) {
        var pages = getCurrentPages();
```

```
        var page = pages[pages.length - 1];
        page.setData({ text: res.tempFilePath });
        wx.getImageInfo({
          src: res.tempFilePath,
          success: function (res) {
            console.log({ text: page.data.text+" width : " + res.width + " height : "
+ res.height });
          }
        })
      }
    })
  },

  /**
   * 生命周期函数--监听页面初次渲染完成
   */
  onReady: function () {

  },

  /**
   * 生命周期函数--监听页面显示
   */
  onShow: function () {

  },

  /**
   * 生命周期函数--监听页面隐藏
   */
  onHide: function () {

  },

  /**
   * 生命周期函数--监听页面卸载
   */
  onUnload: function () {

  },

  /**
   * 页面相关事件处理函数--监听用户下拉动作
   */
  onPullDownRefresh: function () {

  },

  /**
   * 页面上拉触底事件的处理函数
   */
  onReachBottom: function () {

  },

  /**
```

```
   * 用户单击右上角分享
   */
  onShareAppMessage: function () {

  }
})
```

服务端需要部署一张图片，对外访问路径为 http://localhost/wx/wx.jpg。

编译微信小程序后，Console 打印下载的文件地址，如图 6-16 所示。

```
调试器  问题  输出  终端
Wxml  AppData  Console  Sources  Storage  Network  Memory  Security  Sensor  Mock  Audits  Trace
appservice  Filter  Default levels ▼
Some messages have been moved to the Issues panel.
[sitemap 索引情况提示] 根据 sitemap 的规则[0]，当前页面 [pages/index/index] 将被索引
▶{text: "http://tmp/ChNHELOkoAeJ2fee3dff3bd67fe344dda703422d83d4.jpg  width : 960 height : 960"}
>
```

图 6-16

使用微信小程序开发工具可以查看下载的图片信息，如图 6-17 所示。

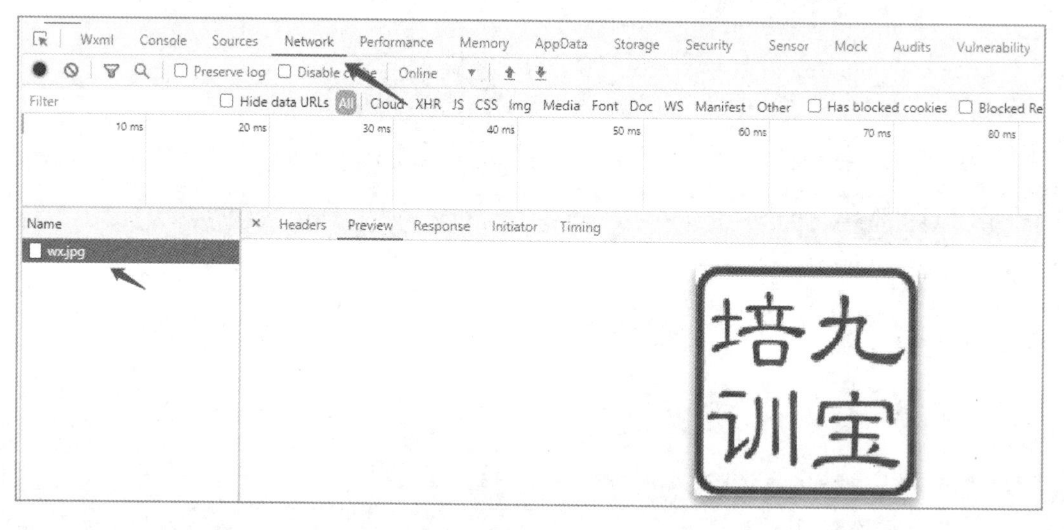

图 6-17

6.2.4 WebSocket

WebSocket 需要同时配置服务端与微信小程序客户端。其实现思路是客户端连接服务端后，服务端与客户端保持“持久连接”，服务端可以向客户端发送数据，亦可以接收客户端传来的数据。

1. 微信小程序端

微信小程序端代码包括 JS 和 WXML 两部分。

（1）JS 程序代码如下。

```
Page({

  /**
```

```
   * 页面的初始数据
   */
  data: {
    text: '九宝老师'
  },
  connectSocket: function () {
    wx.connectSocket({
      url: 'ws://192.168.1.4/wx/WebSocket',
      data: {

      },
      header: {
        'content-type': 'application/json'
      },
      method: "GET",
      success: function () {
        console.log('success')
      },
      fail: function () {
        console.log('fail')
      },
      complete: function () {
        console.log('complete')
      }
    })
  },
  sendSocketMessage: function () {
    wx.sendSocketMessage({
      data: 'jiubao ' + new Date()
    })
  },
  closeSocket: function () {
    wx.closeSocket()
  },

  /**
   * 生命周期函数--监听页面初次渲染完成
   */
  onReady: function () {
    wx.onSocketOpen(function (res) {
      console.log('onSocketOpen')

getCurrentPages()[getCurrentPages().length-1].setData({'text':'onSocketOpen'})
    })
    wx.onSocketError(function (res) {
      console.log("onSocketError   " + res)
      getCurrentPages()[getCurrentPages().length-1].setData({'text':"onSocketError
" + res})
    })
    wx.onSocketMessage(function (res) {
      console.log('onSocketMessage   ' + res.data)

getCurrentPages()[getCurrentPages().length-1].setData({'text':'onSocketMessage   ' +
res.data})
    })
    wx.onSocketClose(function (res) {
```

```
      console.log('onSocketClose')

getCurrentPages()[getCurrentPages().length-1].setData({'text':'onSocketClose'})
     })
   },

   /**
    * 生命周期函数--监听页面加载
    */
   onLoad: function (options) {

   },

   /**
    * 生命周期函数--监听页面显示
    */
   onShow: function () {

   },

   /**
    * 生命周期函数--监听页面隐藏
    */
   onHide: function () {

   },

   /**
    * 生命周期函数--监听页面卸载
    */
   onUnload: function () {

   },

   /**
    * 页面相关事件处理函数--监听用户下拉动作
    */
   onPullDownRefresh: function () {

   },

   /**
    * 页面上拉触底事件的处理函数
    */
   onReachBottom: function () {

   },

   /**
    * 用户单击右上角分享
    */
   onShareAppMessage: function () {

   }
 })
```

（2）WXML 程序代码如下。

```
<view>
    <view>{{text}}</view>
    <button type='warn' bindtap='connectSocket'>connectSocket</button>
    <button type='warn' bindtap='sendSocketMessage'>sendSocketMessage</button>
    <button type='warn' bindtap='closeSocket'>closeSocket</button>
</view>
```

2. 服务端程序

服务端程序代码如下。

```
package wxsocket;

import java.io.IOException;
import java.util.HashSet;
import java.util.Set;

import javax.websocket.OnClose;
import javax.websocket.OnError;
import javax.websocket.OnMessage;
import javax.websocket.OnOpen;
import javax.websocket.Session;
import javax.websocket.server.ServerEndpoint;

@ServerEndpoint(value = "/WebSocket")
public class WebSocket {

public static final Set<WebSocket> connections = new HashSet<WebSocket>();

private Session session;

@OnOpen
public void start(Session session) {
     this.session = session;
     connections.add(this);
     broadcast("@OnOpen");
}

@OnClose
public void end() {
     connections.remove(this);
     broadcast("@OnClose");
}

@OnMessage
public void incoming(String message) {
     broadcast("@OnMessage " + message);
}

@OnError
public void onError(Throwable t) throws Throwable {
     System.out.println("@OnError");
}

private static void broadcast(String msg) {
     for (WebSocket client : connections) {
          try {
```

```
                synchronized (client) {
                    client.session.getBasicRemote().sendText(msg);
                }
            } catch (IOException e) {
                connections.remove(client);
                try {
                    client.session.close();
                } catch (IOException e1) {
                    e1.printStackTrace();
                }
            }
        }
    }

}
```

上述示例程序实现的逻辑：服务端接收 websocket 请求，同时将请求返回给客户端。

测试过程：先启动服务端应用程序，然后编译微信小程序，如图 6-18 所示；单击 connectSocket 按钮，微信小程序将连接服务器，如图 6-19 所示；单击 sendSocketMessage 按钮，发送信息到服务器，如图 6-20 所示。

图 6-18

图 6-19

图 6-20

服务器收到信息后，将信息返回微信小程序客户端，通过数据绑定渲染到显示层，如图 6-21 所示。

WebSocket 关闭前，服务器会主动向微信小程序发送信息，如图 6-22 所示。

图 6-21

图 6-22

调试器 Console 处会打印相关信息，如图 6-23 所示。

图 6-23

单击 closeSocket 按钮，可关闭 WebSocket，如图 6-24 所示。

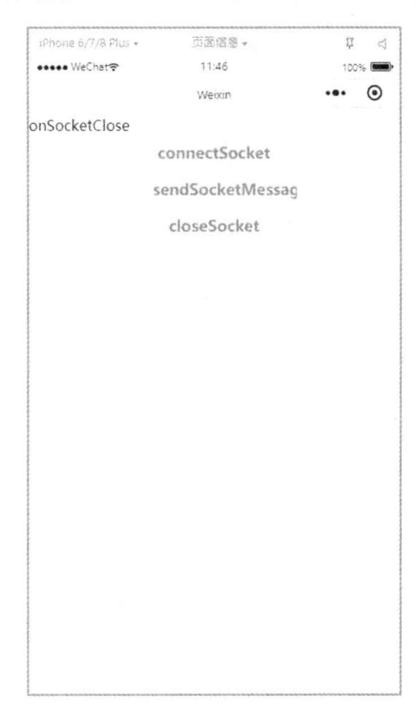

图 6-24

调试器 Console 处会打印关闭 WebSocket 信息，如图 6-25 所示。

```
调试器　问题　输出　终端
Wxml　AppData　Console　Sources　Storage　Network　Memory　Security　Sensor　Mock　Audits　Trace　Vulnerability
appservice　Filter　Default levels
Some messages have been moved to the Issues panel.
[sitemap 索引情况提示] 根据 sitemap 的规则[0]，当前页面 [pages/index/index] 将被索引
success
complete
Sun May 30 2021 11:46:28 GMT+0800 (中国标准时间) 配置中关闭合法域名、web-view (业务域名) 、TLS 版本以及 HTTPS 证书检查
工具未校验合法域名、web-view (业务域名) 、TLS 版本以及 HTTPS 证书。
onSocketOpen
onSocketMessage　@OnOpen
onSocketMessage　@OnMessage jiubao Sun May 30 2021 11:46:29 GMT+0800 (中国标准时间)
onSocketMessage　@OnMessage 服务器主动发送信息
onSocketClose
```

图 6-25

6.3 数据 API

使用微信小程序时，用户的数据会被存储在本地缓存指定的 key 中，覆盖原来该 key 对应的内容。除非用户主动删除数据，或因系统存储空间过小而被清理，否则数据将一直可用。

单个 key 允许存储的最大数据长度为 1 MB，所有数据的存储上限为 10 MB。

注意：

虽然 localStorage 可以永久存储数据，但不建议将关键信息都存储在 localStorage 中，以避免用户更换设备时产生数据丢失或外泄问题。

数据 API 的重点同样是同步和异步问题。此外，微信小程序进行数据的定义（set 方法）、获取（get 方法）等操作时相对简单。下面来看一段示例程序。

（1）JS 程序代码如下。

```
Page({

  /**
   * 页面的初始数据
   */
  data: {

  },
  setStorage: function () {
    wx.setStorage({
      key: "jiubao",
      data: "九宝"
    })
  },
  setStorageSync: function () {
    wx.setStorageSync('jiubao', '九宝')
  },

  getStorage: function () {
    wx.getStorage({
      key: 'jiubao',
      success: function (res) {
        console.log(res.data)
      }
    })
  },
  getStorageSync: function () {
    var value = wx.getStorageSync('jiubao')
    console.log(value);
  },

  getStorageInfo: function () {
    wx.getStorageInfo({
      success: function (res) {
        console.log(res.keys)
        console.log(res.currentSize)
        console.log(res.limitSize)
      }
    })
  },
  getStorageInfoSync: function () {
```

```
    var res = wx.getStorageInfoSync()
    console.log(res.keys)
    console.log(res.currentSize)
    console.log(res.limitSize)
  },

  removeStorage: function () {
    wx.removeStorage({
      key: 'jiubao',
      success: function (res) {
        console.log(res.data)
      }
    })
  },
  removeStorageSync: function () {
    wx.removeStorageSync('jiubao')
  },

  clearStorage: function () {
    wx.clearStorage()
  },
  clearStorageSync: function () {
    wx.clearStorageSync()
  },

  /**
   * 生命周期函数--监听页面初次渲染完成
   */
  onReady: function () {
    wx.onSocketOpen(function (res) {
      console.log('onSocketOpen')

getCurrentPages()[getCurrentPages().length-1].setData({'text':'onSocketOpen'})
    })
    wx.onSocketError(function (res) {
      console.log("onSocketError  " + res)
      getCurrentPages()[getCurrentPages().length-1].setData({'text':"onSocketError
" + res})
    })
    wx.onSocketMessage(function (res) {
      console.log('onSocketMessage  ' + res.data)

getCurrentPages()[getCurrentPages().length-1].setData({'text':'onSocketMessage  ' +
res.data})
    })
    wx.onSocketClose(function (res) {
      console.log('onSocketClose')

getCurrentPages()[getCurrentPages().length-1].setData({'text':'onSocketClose'})
    })
  },

  /**
   * 生命周期函数--监听页面加载
   */
  onLoad: function (options) {

  },
```

```
    /**
     * 生命周期函数--监听页面显示
     */
    onShow: function () {

    },

    /**
     * 生命周期函数--监听页面隐藏
     */
    onHide: function () {

    },

    /**
     * 生命周期函数--监听页面卸载
     */
    onUnload: function () {

    },

    /**
     * 页面相关事件处理函数--监听用户下拉动作
     */
    onPullDownRefresh: function () {

    },

    /**
     * 页面上拉触底事件的处理函数
     */
    onReachBottom: function () {

    },

    /**
     * 用户单击右上角分享
     */
    onShareAppMessage: function () {

    }
  })
```

（2）WXML 文件代码如下。

```
    <view>
        <view class='but' type='warn' bindtap='setStorage'>setStorage</view>
        <view class='butSync' type='primary'
bindtap='setStorageSync'>setStorageSync</view>
    </view>
    <view>
        <view class='but' type='warn' bindtap='getStorage'>getStorage</view>
        <view class='butSync' type='primary'
bindtap='getStorageSync'>getStorageSync</view>
    </view>
    <view>
        <view class='but' type='warn' bindtap='getStorageInfo'>getStorageInfo</view>
        <view class='butSync' type='primary'
bindtap='getStorageInfoSync'>getStorageInfoSync</view>
```

```
    </view>
    <view>
        <view class='but' type='warn' bindtap='removeStorage'>removeStorage</view>
        <view class='butSync' type='primary'
bindtap='removeStorageSync'>removeStorageSync</view>
    </view>
    <view>
        <view class='but' type='warn' bindtap='clearStorage'>clearStorage</view>
        <view class='butSync' type='primary'
bindtap='clearStorageSync'>clearStorageSync</view>
    </view>
```

（3）WXSS 文件代码如下。

```
    view{
      margin: 50rpx;
    }

    .but{
      text-align: center;
      background-color: green;
      color: white;
      border: 1px black solid;
      border-radius: 15rpx;
      margin: 15rpx;
    }

    .butSync{
      text-align: center;
      background-color: red;
      color: white;
      border: 1px black solid;
      border-radius: 15rpx;
      margin: 15rpx;
    }
```

编译微信小程序，效果如图 6-26 所示。

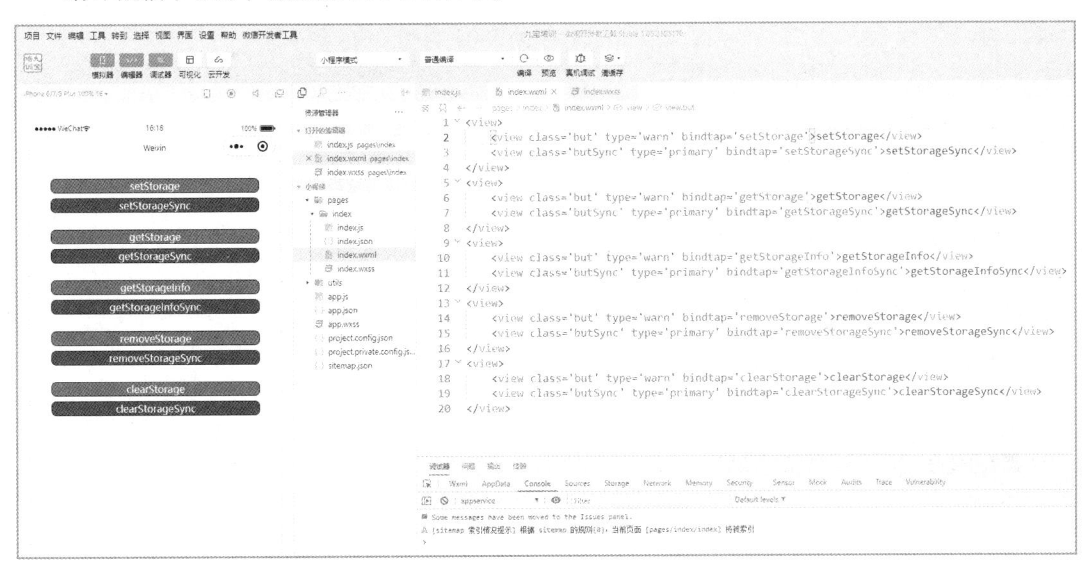

图 6-26

wx.setStorage 与 wx.setStorageSync 都用于将数据存储在本地缓存指定的 key 中。单击 setStorage 与 setStorageSync 按钮，即可将“jiubao：九宝”保存，如图 6-27 所示。

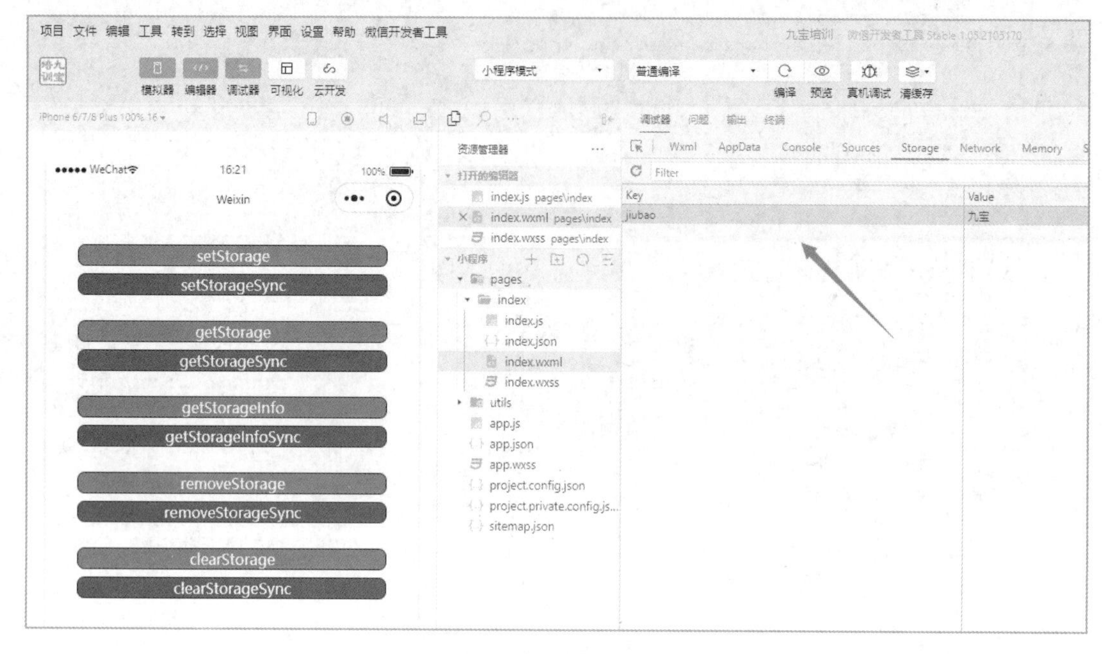

图 6-27

wx.removeStorage 与 wx.removeStorageSync 都用于从本地缓存指定的 key 中删除数据。单击 removeStorage 与 removeStorageSync 按钮，即可将 key 是 jiubao 的数据删除，如图 6-28 所示。

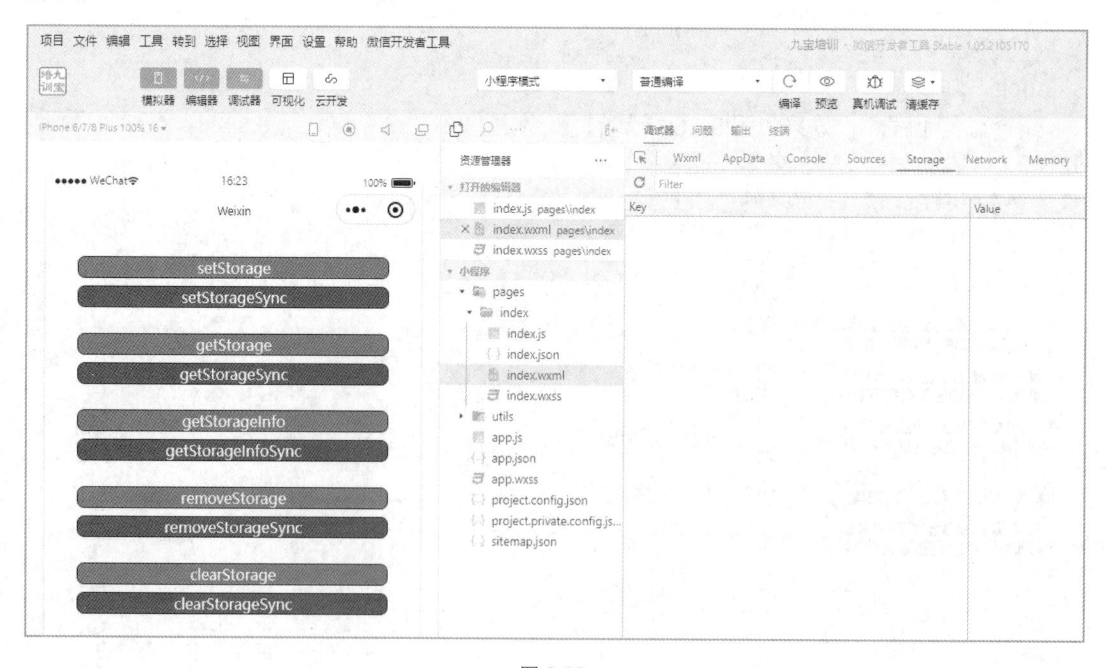

图 6-28

wx.getStorage 与 wx.getStorageSync 都用于获取 storage 的相关信息 。单击 getStorage

与 getStorageSync 按钮，即可将 storage 的相关信息输出到 Console，如图 6-29 所示。

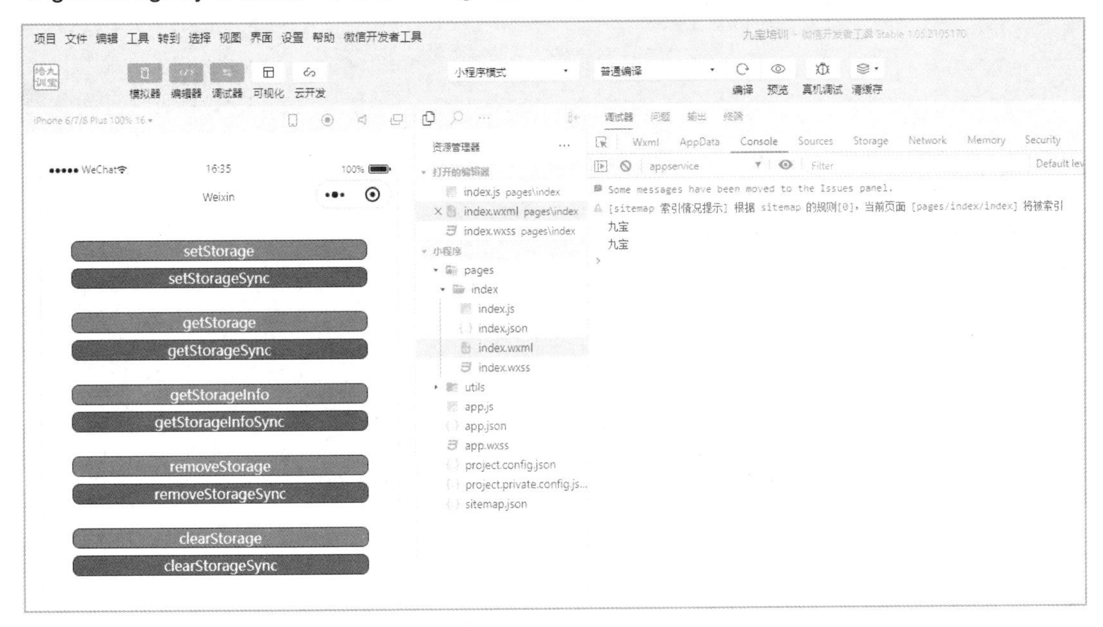

图 6-29

wx.getStorageInfo 与 wx.getStorageInfoSync 都用于获取 storage 的相关信息。单击 getStorageInfo 与 getStorageInfoSync 按钮，即可将 storage 的相关信息输出到 Console，如图 6-30 所示。

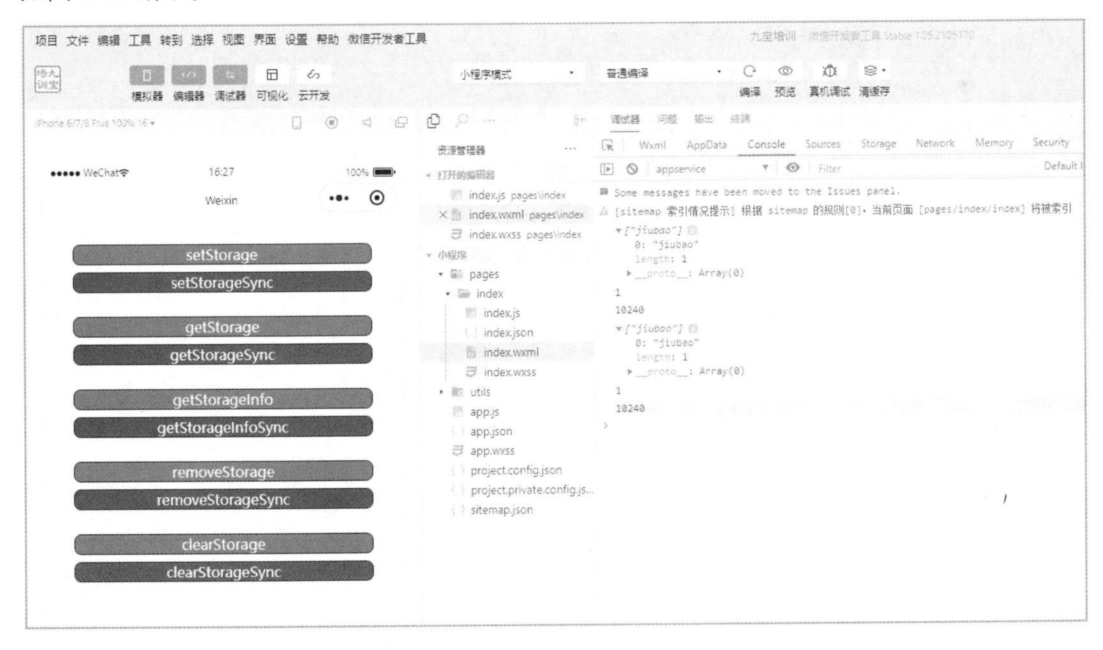

图 6-30

wx.clearStorage 与 wx.clearStorageSync 都用于清理本地数据缓存。单击 clearStorage 与 clearStorageSync 按钮，即可将数据清除，如图 6-31 所示。

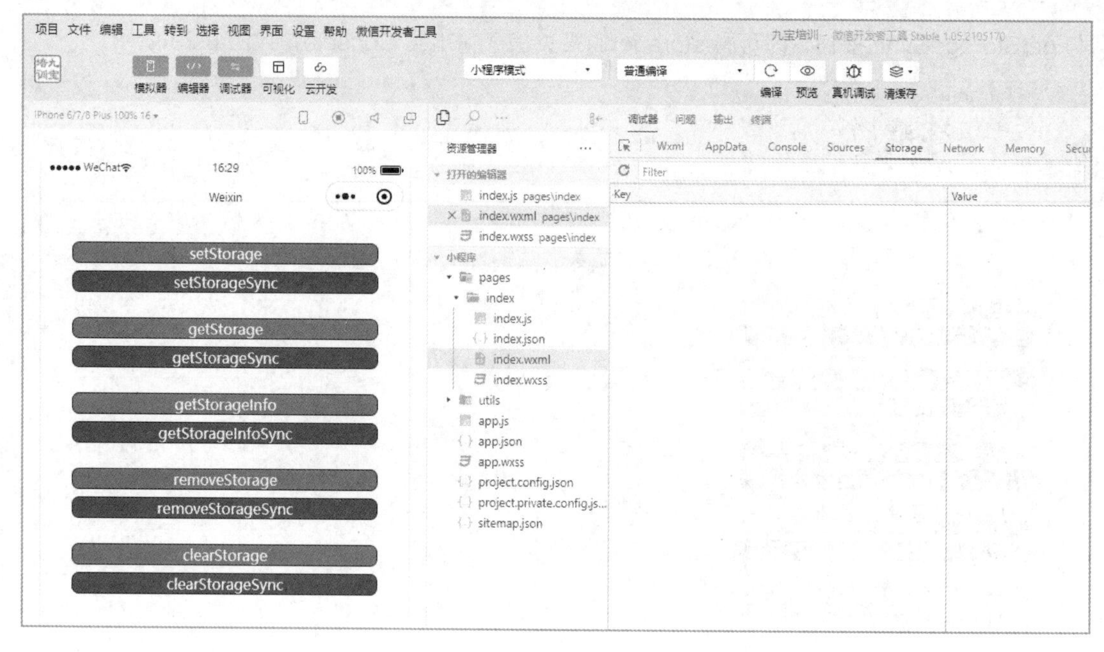

图 6-31

6.4 位置 API

位置 API 用于实现与位置有关的功能，下面来看示例程序。

（1）JS 文件代码如下。

```
Page({

  /**
   * 页面的初始数据
   */
  data: {
    latitude:'',
    longitude:'',
  },
  chooseLocation: function () {
    wx.chooseLocation({success:function(res){
      console.log('name :'+res.name);
      console.log('address :'+res.address);
      console.log('latitude :'+res.latitude);
      console.log('longitude :'+res.longitude);

getCurrentPages()[getCurrentPages().length-1].setData({latitude:res.latitude})

getCurrentPages()[getCurrentPages().length-1].setData({longitude:res.longitude})
    }});
  },
  openLocation: function () {
    wx.openLocation({
      latitude: this.data.latitude-0,
      longitude: this.data.longitude-0,
```

```
      scale: 28
    })
  },

  /**
   * 生命周期函数--监听页面初次渲染完成
   */
  onReady: function () {
    wx.onSocketOpen(function (res) {
      console.log('onSocketOpen')

getCurrentPages()[getCurrentPages().length-1].setData({'text':'onSocketOpen'})
    })
    wx.onSocketError(function (res) {
      console.log("onSocketError  " + res)
      getCurrentPages()[getCurrentPages().length-1].setData({'text':"onSocketError
" + res})
    })
    wx.onSocketMessage(function (res) {
      console.log('onSocketMessage  ' + res.data)

getCurrentPages()[getCurrentPages().length-1].setData({'text':'onSocketMessage  ' +
res.data})
    })
    wx.onSocketClose(function (res) {
      console.log('onSocketClose')

getCurrentPages()[getCurrentPages().length-1].setData({'text':'onSocketClose'})
    })
  },

  /**
   * 生命周期函数--监听页面加载
   */
  onLoad: function (options) {

  },

  /**
   * 生命周期函数--监听页面显示
   */
  onShow: function () {

  },

  /**
   * 生命周期函数--监听页面隐藏
   */
  onHide: function () {

  },

  /**
   * 生命周期函数--监听页面卸载
   */
  onUnload: function () {
```

```
  },

  /**
   * 页面相关事件处理函数--监听用户下拉动作
   */
  onPullDownRefresh: function () {

  },

  /**
   * 页面上拉触底事件的处理函数
   */
  onReachBottom: function () {

  },

  /**
   * 用户单击右上角分享
   */
  onShareAppMessage: function () {

  }
})
```

（2）WXML 文件代码如下。

```
<view>
    <view>{{text}}</view>
    <button type='warn' bindtap='chooseLocation'>chooseLocation</button>
    <button type='warn' bindtap='openLocation'>openLocation</button>
</view>
```

程序编译后效果如图 6-32 所示。

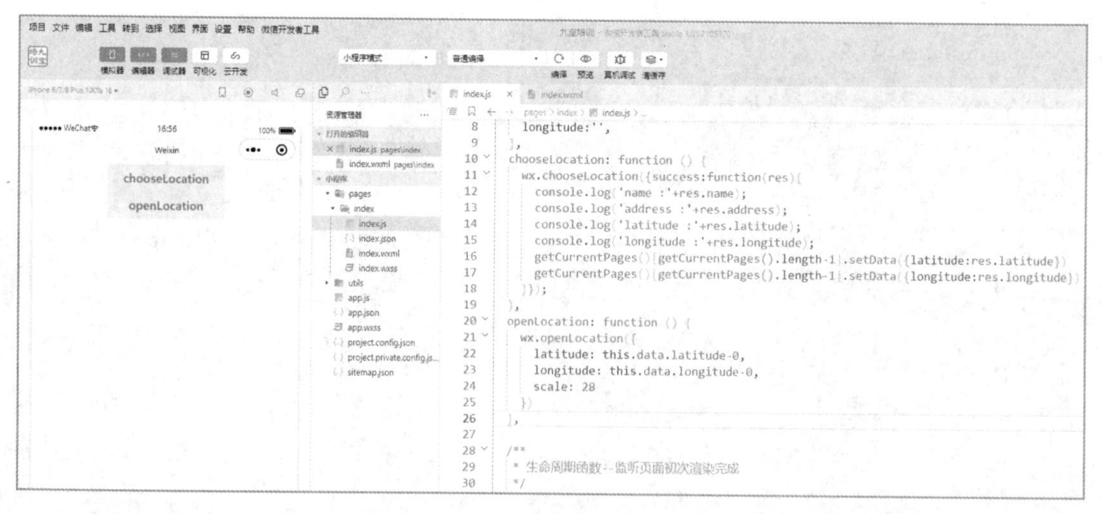

图 6-32

在模拟器中单击 chooseLocation 按钮，将给出当前位置提示，如图 6-33 所示。

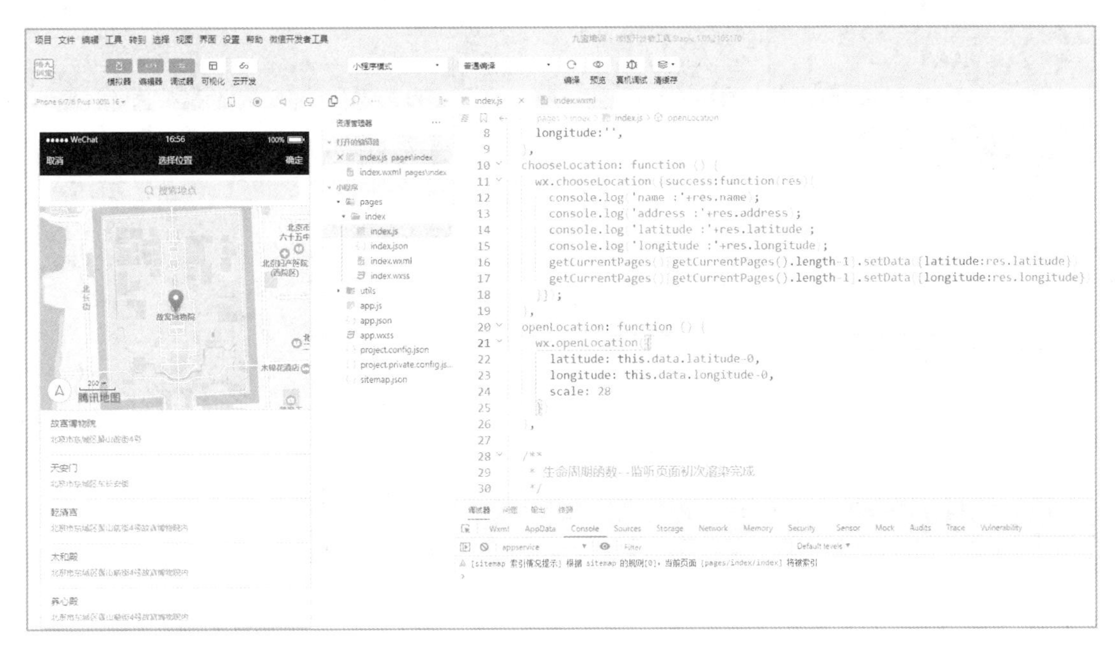

图 6-33

选择具体的地址后，调试器中会显示相关信息，如图 6-34 所示。

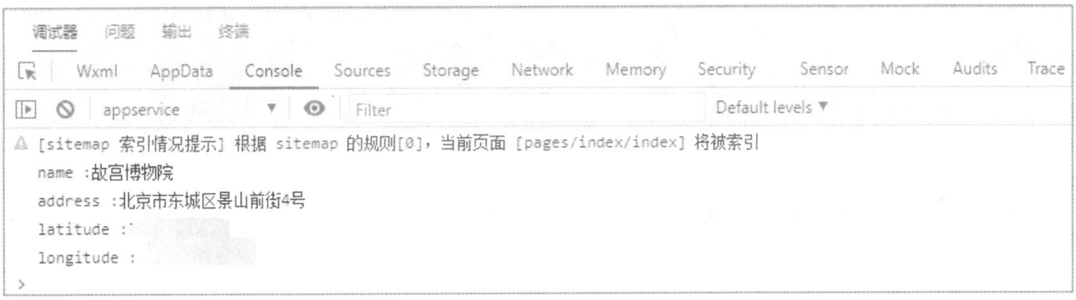

调试器　问题　输出　终端

Wxml　AppData　Console　Sources　Storage　Network　Memory　Security　Sensor　Mock　Audits　Trace

appservice　Filter　Default levels

```
[sitemap 索引情况提示] 根据 sitemap 的规则[0]，当前页面 [pages/index/index] 将被索引
name :故宫博物院
address :北京市东城区景山前街4号
latitude :
longitude :
```

图 6-34

微信小程序会保存本地坐标信息，如图 6-35 所示。

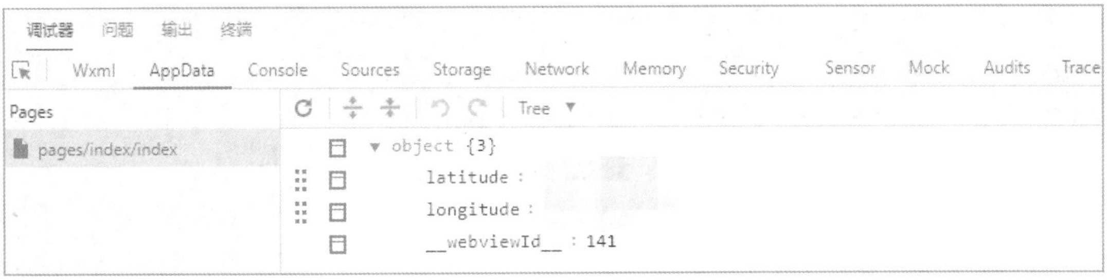

调试器　问题　输出　终端

Wxml　AppData　Console　Sources　Storage　Network　Memory　Security　Sensor　Mock　Audits　Trace

Pages　Tree

pages/index/index

```
object {3}
    latitude :
    longitude :
    __webviewId__ : 141
```

图 6-35

在模拟器中单击 openLocation 按钮，可发现地图显示的就是刚才选择的地址，如图 6-36 所示。

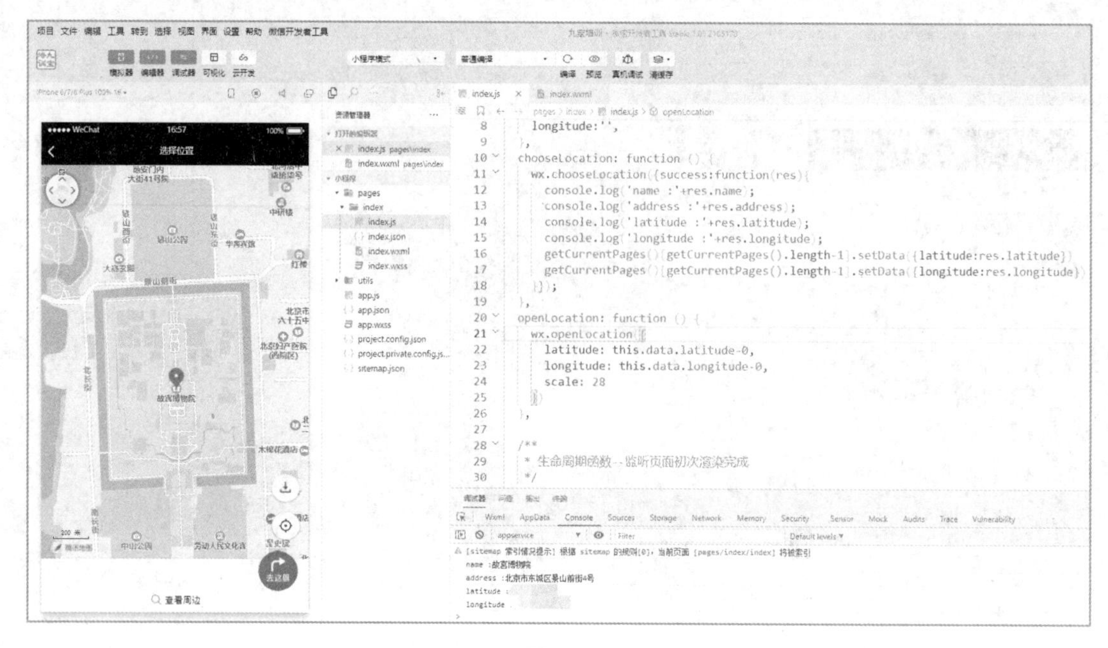

图 6-36

6.5 设备 API

下面以获取网络类型、系统信息、设备电量为例，介绍微信小程序的设备 API。

（1）JS 文件代码如下。

```
Page({

  /**
   * 页面的初始数据
   */
  data: {
    text: '九宝'
  },

  getNetworkType: function () {
    wx.getNetworkType({
      success: function (res) {
        var networkType = res.networkType // 返回网络类型 2g，3g，4g，Wi-Fi
        console.log(networkType);
      }
    })
  },
  getSystemInfo: function () {
    wx.getSystemInfo({
      success: function (res) {
        console.log(res.model)
        console.log(res.pixelRatio)
        console.log(res.windowWidth)
        console.log(res.windowHeight)
        console.log(res.language)
```

```
        console.log(res.version)
      }
    })
  },
  makePhoneCall: function () {
    wx.makePhoneCall({
      phoneNumber: '1300000' //仅为示例，并非真实的电话号码
    })
  },

  /**
   * 生命周期函数--监听页面加载
   */
  onLoad: function (options) {

  },

  /**
   * 生命周期函数--监听页面初次渲染完成
   */
  onReady: function () {

  },

  /**
   * 生命周期函数--监听页面显示
   */
  onShow: function () {

  },

  /**
   * 生命周期函数--监听页面隐藏
   */
  onHide: function () {

  },

  /**
   * 生命周期函数--监听页面卸载
   */
  onUnload: function () {

  },

  /**
   * 页面相关事件处理函数--监听用户下拉动作
   */
  onPullDownRefresh: function () {

  },

  /**
   * 页面上拉触底事件的处理函数
   */
  onReachBottom: function () {
```

```
  },

  /**
   * 用户单击右上角分享
   */
  onShareAppMessage: function () {

  }
})
```

（2）WXML 文件代码如下。

```
<view>
    <view>{{text}}</view>
    <view>{{str}}</view>
    <button bindtap='getNetworkType'>getNetworkType</button>
    <button bindtap='getSystemInfo'>getSystemInfo</button>
    <button bindtap='getBatteryInfo'>getBatteryInfo</button>
</view>
```

程序编译后效果如图 6-37 所示。

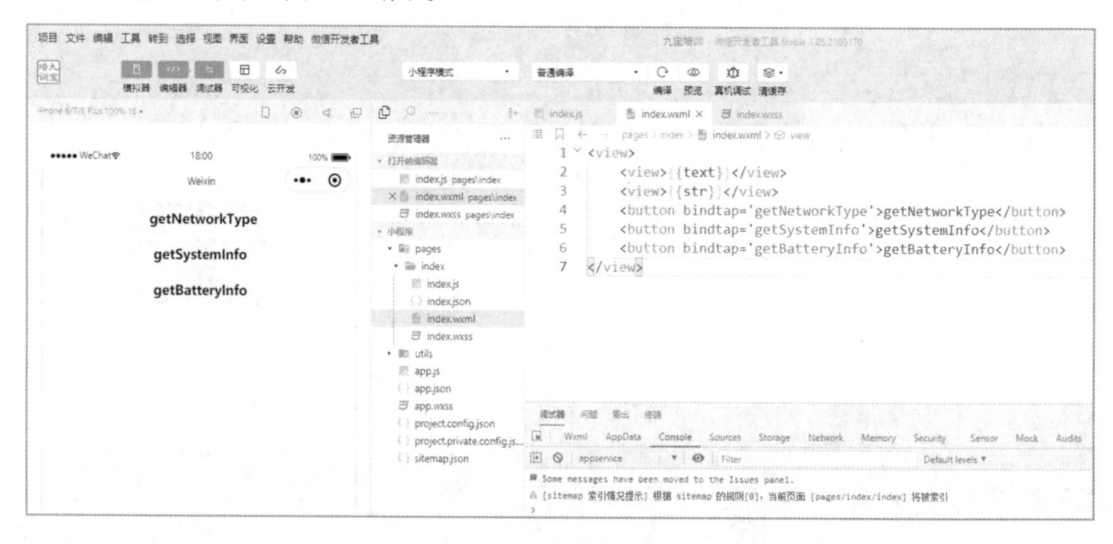

图 6-37

在模拟器中单击 getNetworkType 按钮，调试器中的信息如图 6-38 所示。

```
调试器  问题  输出  终端
Wxml  AppData  Console  Sources  Storage  Network  Memory  Security  Sensor  Mock  Audits  Trace
appservice  Filter  Default levels
Some messages have been moved to the Issues panel.
[sitemap 索引情况提示] 根据 sitemap 的规则[0]，当前页面 [pages/index/index] 将被索引
▼{errMsg: "getNetworkType:ok", networkType: "wifi"}
   errMsg: "getNetworkType:ok"
   networkType: "wifi"
 ▶__proto__: Object
>
```

图 6-38

单击 getSystemInfo 按钮，调试器中的信息如图 6-39 所示。

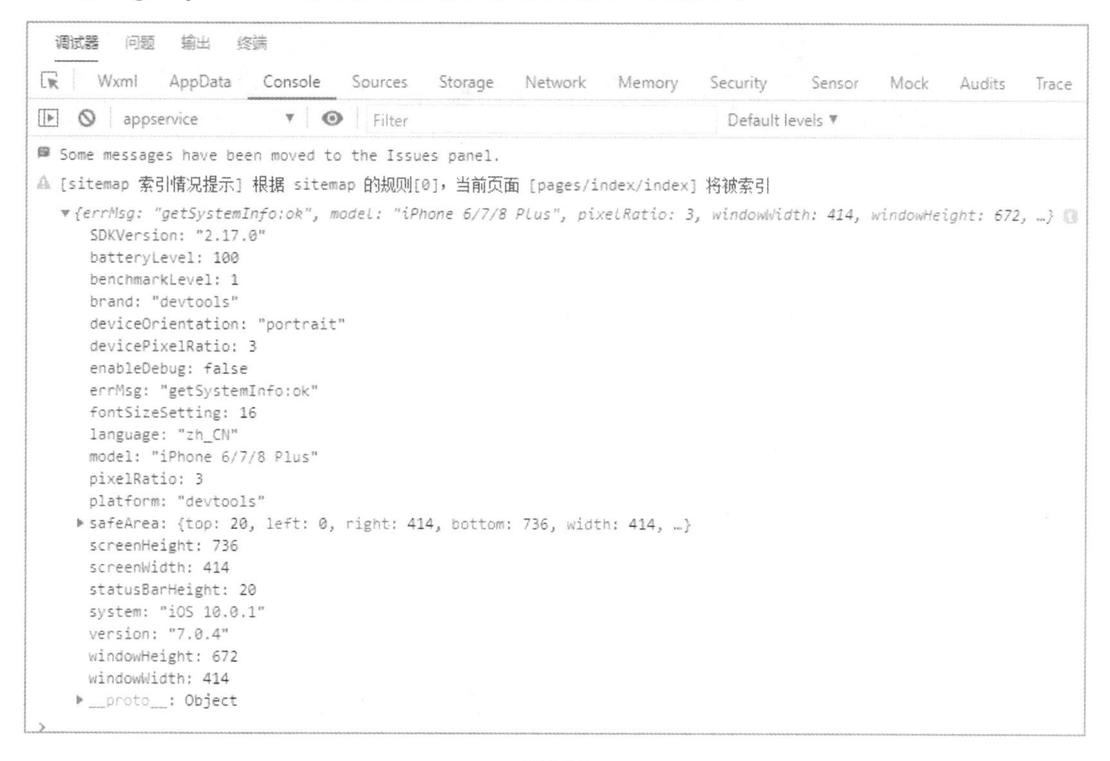

图 6-39

单击 getBatteryInfo 按钮，调试器中的信息如图 6-40 所示。

图 6-40

第7章　小程序支付

本章将详细讲解微信小程序支付的相关知识点。

读者需注意，本章讲解的微信小程序开发需要服务端程序与微信小程序客户端程序配合使用。因此，本章将给出微信小程序中微信支付所需的示例程序。

7.1　微信小程序支付相关知识点

微信小程序支付相关知识点如下。

（1）需要明确微信小程序与微信支付之间的关系。从技术角度讲，微信小程序与微信支付是两个独立的系统，不是“包含”的关系。一般来说，微信小程序申请认证通过后，可以申请微信支付，审核通过后即可开始开发。到本书截稿为止，微信支付尚不提供“开发测试账号”。

（2）微信支付有 V2、V3 两个版本，本书基于最新微信支付版本 V3 展开讲解。V2 版接口和 V3 版接口是基于两种接口标准设计的两套接口，目前实际应用中的大部分接口均已升级为 V3 版接口。

V2 和 V3 两个接口版本的区别如表 7-1 所示。

表 7-1

规则差异	V2 版本	V3 版本
参数格式	XML	JSON
提交方式	POST	POST、GET、PATCH、DELETE
回调加密	无须加密	AES-256-GCM 加密
敏感加密	无须加密	RSA 加密
编码方式	UTF-8	UTF-8
签名方式	MD5 或 HMAC-SHA256	非对称密钥 SHA256-RSA

（3）微信支付 V2、V3 的主要差异集中在报文格式与加密、解密算法方面。

（4）微信支付分为直连方式与服务商方式，其业务逻辑大致相同，差异主要体现在相关报文上。本章所讲述的微信支付直连方式的相关算法和业务逻辑适用于服务商方式。

7.2　开发步骤

微信小程序支付通常不是独立的项目，而是某个应用开发的一部分。为了最大限度地降低读者的学习门槛，本章将只实现微信小程序支付功能。

读者学习本章时，可以单独建立一个项目，只实现微信支付功能。当所有技术点都调试正常后，再将相关程序集成到最终的项目中。

微信小程序支付功能的开发一共分为如下 4 步。

（1）获取 openid。

（2）调用“统一下单 API”，以获取 prepay_id。

（3）再次签名。

（4）在微信小程序端调用微信支付功能。

下面我们就按照以上开发顺序，一起来学习如何开发微信小程序支付模块。

7.2.1　获取 openid

微信小程序中，openid 是用户在直连商户 appid 下的唯一标识，是微信支付报文的必需字段。下面首先来获取微信小程序的 openid。

构建项目的“最小程序状态”，并删除不必要的 js 函数。在 JS 文件中定义 4 个函数，参考代码如下，开发页面如图 7-1 所示。

```
step1:function(){

},
step2:function(){

},
step3:function(){

},
step4:function(){

}
```

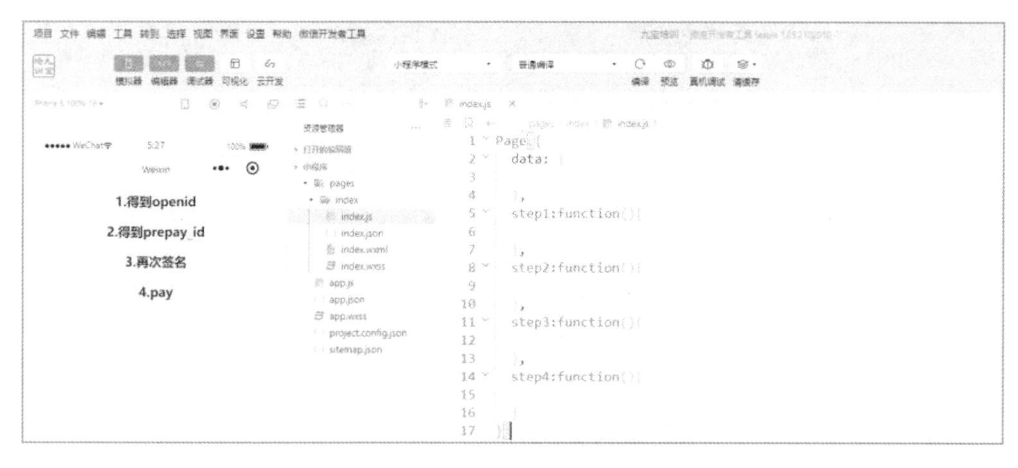

图 7-1

在 WXML 文件中定义 4 个 button 组件，分别对应微信小程序支付的 4 个步骤，代码如下，开发界面如图 7-2 所示。

```
<button bindtap="step1">1.得到 openid</button>
<view>{{bean1.openid}}</view>
<button bindtap="step2">2.得到 prepay_id</button>
<view>{{bean2.prepay_id}}</view>
<button bindtap="step3">3.再次签名</button>
```

```
<view>{{bean3.paySign}}</view>
<button bindtap="step4">4.pay</button>
```

图 7-2

1. 调用 wx.login

调用微信小程序的 wx.login 文件，先获取登录凭证（code），然后通过 code 获取用户的登录状态信息，包括用户在当前小程序中的唯一标识（openid）、微信开放平台账号下的唯一标识（unionid，若当前小程序已绑定到微信开放平台账号）以及本次登录的会话密钥（session_key）等。

修改 JS 文件，调用 wx.login 文件，参考代码如下，开发页面如图 7-3 所示。

```
wx.login({
     success (res) {
         console.log(res.code);
     }
})
```

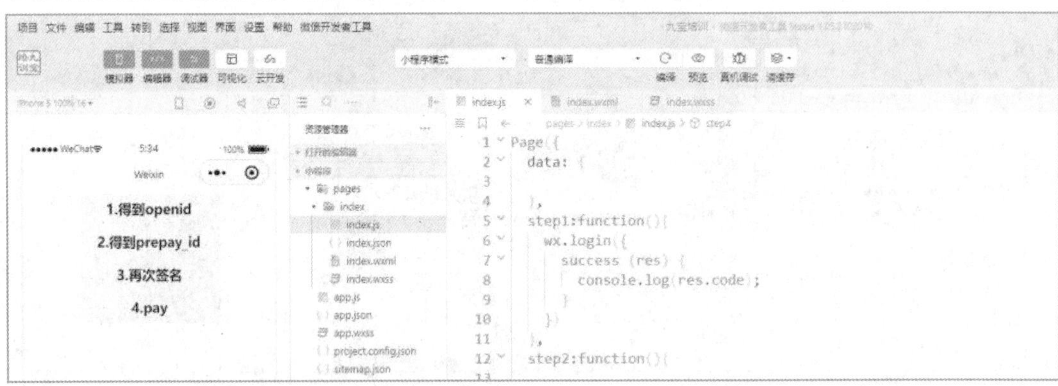

图 7-3

程序执行编译后，在模拟器中单击“1.得到 openid”按钮，调试器中将显示如图 7-4 所示的 code 信息。

图 7-4

2. 调用 wx.request

根据 code 获取 openid 是在服务端完成的，因此需要将获取的 code 提交给服务器。

3. 创建空的 Web Server 项目

本例采用 Java 作为 Server 开发语言，使用 Tomcat 作为 Web Server。为了降低学习门槛，这里不使用 SpringMVC、SpringBoot 等框架工具，只使用最基础的 servlet 来实现相关需要。

使用 servlet 创建项目后，需要完成相关服务配置，以保证本地服务能正常工作。

新建 servlet，示例代码如下。

```
package jiubao2326321088;

import java.io.IOException;
import javax.servlet.ServletException;
import javax.servlet.annotation.WebServlet;
import javax.servlet.http.HttpServlet;
import javax.servlet.http.HttpServletRequest;
import javax.servlet.http.HttpServletResponse;

import jiubao2326321088.bean.BeanStep1;
import jiubao2326321088.util.WXUtil;

@WebServlet("/Step1")
public class Step1 extends HttpServlet {
    protected void doGet(HttpServletRequest request, HttpServletResponse response)
throws ServletException, IOException {
        String code = request.getParameter("code");
        System.out.println("step1");
        response.getWriter().print(WXUtil.getopenid(code));
    }
}
```

doGet 方法用来接收 code 参数。调用 WXUtil.getopenid 方法进行登录凭证校验，相关示例程序如下。

```
public static String getopenid(String js_code){
    try {
        String str = Request.Get("https://api.weixin.qq.com/sns/jscode2session?
appid=" + WX_Args.appid + "&secret=" + WX_Args.secret + "&js_code=" + js_code +
"&grant_type=authorization_code")
                .execute()
                .returnContent()
                .asString();
        return str;
    } catch (Exception e) {
        e.printStackTrace();
        return null;
    }
}
```

public static String getopenid(String js_code)是 WXUtil 类中的方法，其请求参数、返回值如表 7-2、表 7-3 所示。

表 7-2

属　性	类　型	必　填	说　明
appid	string	是	小程序 AppID
secret	string	是	小程序 AppSecret
js_code	string	是	登录时获取的 code
grant_type	string	是	授权类型，此处只需填写 authorization_code

表 7-3

属　性	类　型	说　明
openid	string	用户唯一标识
session_key	string	会话密钥
unionid	string	用户在开放平台的唯一标识符，在满足 UnionID 下发条件的情况下会返回
errcode	number	错误码
errmsg	string	错误信息

注意：

public static String getopenid(String js_code) 只需要返回腾讯服务器响应报文即可。

需要提供的账号信息参见 WX_Args 类，具体代码如下。读者可修改为自己的账号信息。

```
package jiubao2326321088.util;

public class WX_Args {
    public static String appid = "";
    public static String secret = "";
    public static String mchid = "";
    public static String key = "jiubao2326321088jiubao2326321088";
    public static String serial_no = "";
    public static String file = "E:\\九宝培训\\1434507702_20210125_cert\\
apiclient_key.pem";
}
```

代码解析如下。

（1）appid 就是微信小程序的 ID（AppID），secret 是小程序密钥（AppSecret）。在“开发管理”页面的“开发设置”选项卡下可以获取 AppID 和 AppSecret，如图 7-5 所示。

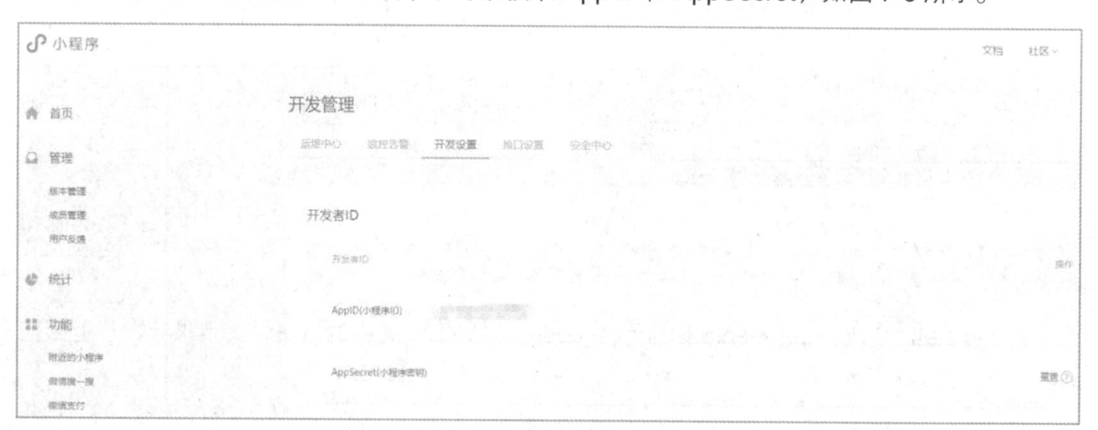

图 7-5

（2）mchid 是微信支付的商户号，申请微信支付时，腾讯回复的邮件中会包含该信息，读者可以在微信商户后台查询。

（3）key 是微信支付密钥，在微信支付回调通知验证请求时需要使用。密钥可在如图 7-6 所示位置进行设置和查询。

账户设置
商户信息
发票信息
API安全
审核配置
微信群管理
员工账号管理
微信商户终端
微信收款商业版
支付申请
消费者投诉
违约记录

API密钥

在API调用时用来按照指定规则对你的请求参数进行签名，服务器收到你的请求时会进行签名验证，既可以界定你的身份也可以防止其他人恶意篡改请求数据。
API密钥的设置与修改不影响APIv3密钥

1 安装操作证书
API密钥属于敏感信息，只有安装了操作证书的用户才能操作
你已安装操作证书
证书管理

2 设置API密钥
API密钥属于敏感信息，请妥善保管不要泄露，如果怀疑信息泄露，请重设密钥。
设置密钥

APIv3密钥

调用APIv3的下载平台证书接口、处理回调通知中报文时，要通过该密钥来解密信息，防止报文被他人恶意篡改。 查看帮助
APIv3密钥的设置与修改不影响API密钥

设置APIv3密钥
你已于2021年01月25日09:12成功设置密钥
你已设置密钥
重设密钥

图 7-6

（4）serial_no 是 API 证书编号，在“账户中心”页面的“API 安全”选项卡下可以获取 API 证书编号，如图 7-7、图 7-8 所示。

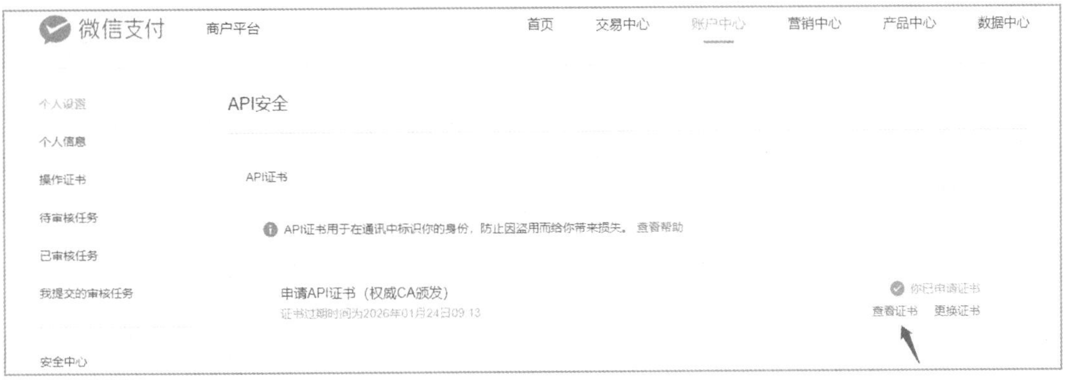

微信支付　商户平台　首页　交易中心　账户中心　营销中心　产品中心　数据中心

个人设置
个人信息
操作证书
待审核任务
已审核任务
我提交的审核任务
安全中心

API安全

API证书

API证书用于在通讯中标识你的身份，防止因盗用而给你带来损失。 查看帮助

申请API证书（权威CA颁发）
证书过期时间为2026年01月24日09:13
你已申请证书
查看证书　更换证书

图 7-7

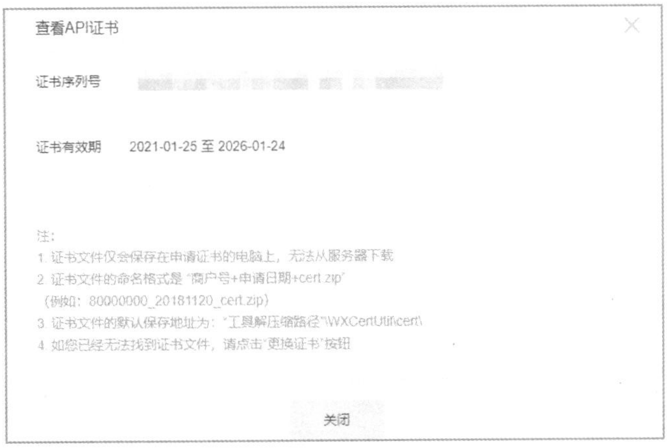

查看API证书

证书序列号

证书有效期　2021-01-25 至 2026-01-24

注：
1. 证书文件仅会保存在申请证书的电脑上，无法从服务器下载
2. 证书文件的命名格式是“商户号+申请日期+cert.zip”
（例如：80000000_20181120_cert.zip）
3. 证书文件的默认保存地址为：“工具解压缩路径”\WXCertUtil\cert\
4. 如您已经无法找到证书文件，请点击“更换证书”按钮

关闭

图 7-8

4. 申请微信支付证书

该操作的步骤较多，需要先登录微信支付后台，在“API 安全”栏的“API 证书管理”区域下载证书下载工具，然后使用证书下载工具申请证书，如图 7-9 ~ 图 7-15 所示，按照操作向导单击“下一步”即可。

图 7-9

图 7-10

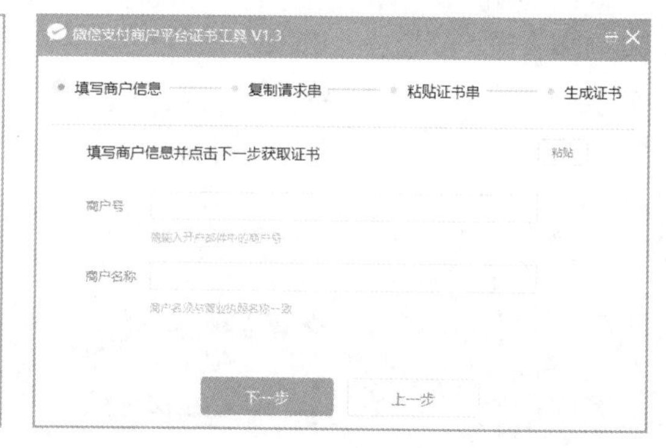

图 7-11

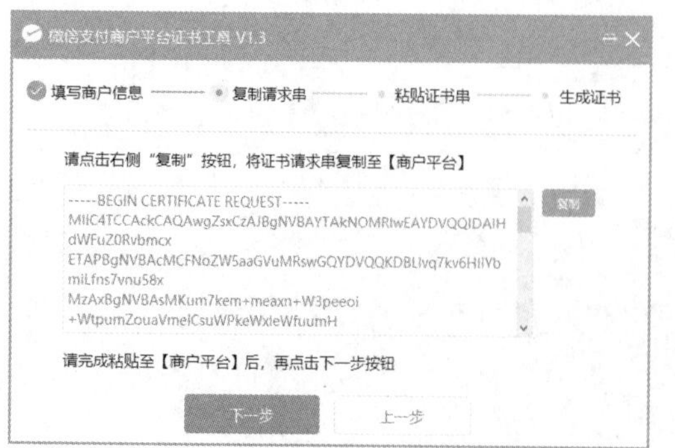

图 7-12

图 7-13

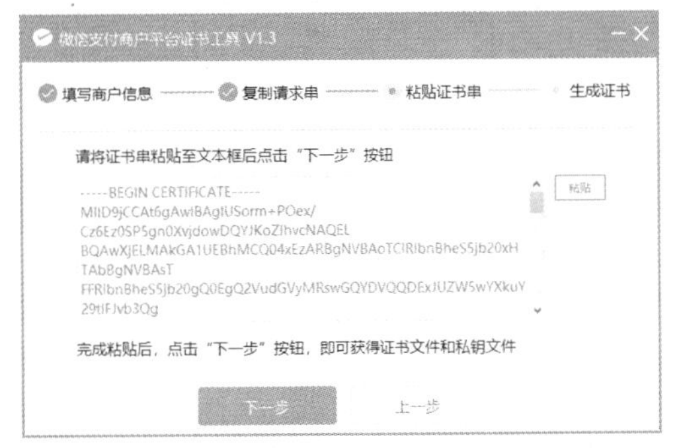

图 7-14

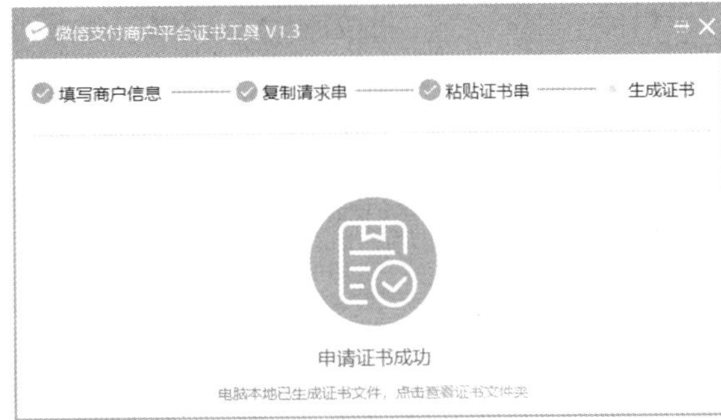

图 7-15

5. 微信小程序端发起请求

微信小程序端获取 code 后，即可发起网络请求，相关示例程序如下。

```
wx.request({
     url: 'http://localhost/wpay3_xcx/Step1',
     data: {
          code: res.code
     },
     success:function(res){
          var pages = getCurrentPages();
          var page = pages[pages.length-1];
          page.setData({bean1:res.data});
     }
})
```

注意：

（1）当前环境为开发环境，因此网络请求未验证域名。但仍需注意微信小程序开发工具的相关配置。

（2）微信小程序 wx.request 是在 wx.login 请求成功的情况下才能调用的，对于 wx.login 请求失败的情况，读者可自行酌情解决。

（3）微信小程序 wx.request 的 success 响应是异步的，success 保存 openid 等信息需要获取当前 page。考虑到实际编程情况，本小节使用 getCurrentPages()获取当前页面栈后，可以得到当前 page。同样，对于 wx.request 请求失败的情况，读者可自行酌情解决，如图 7-16 所示。

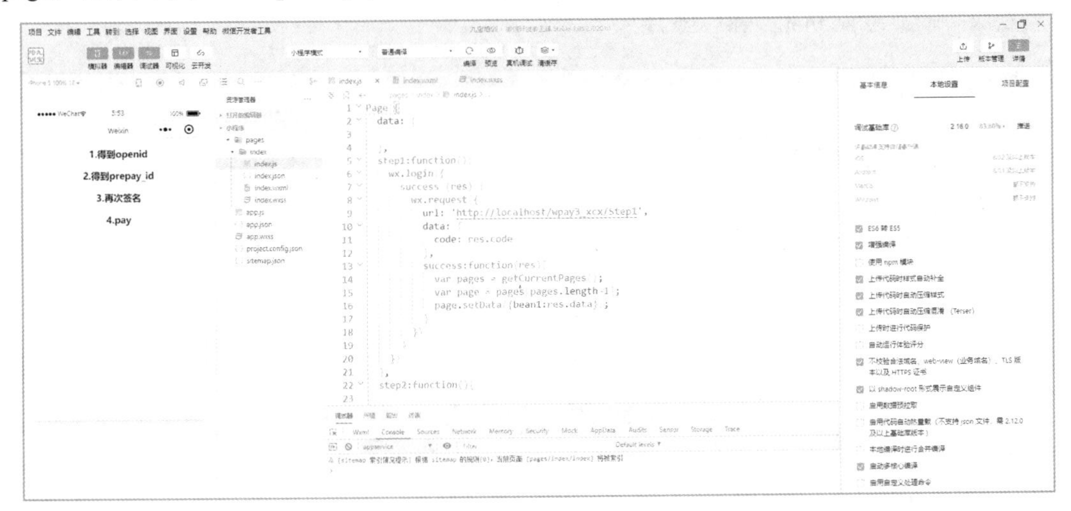

图 7-16

完成上述工作后，编译微信小程序，单击“1.得到 openid”按钮，即可得到相关信息，如图 7-17 所示。

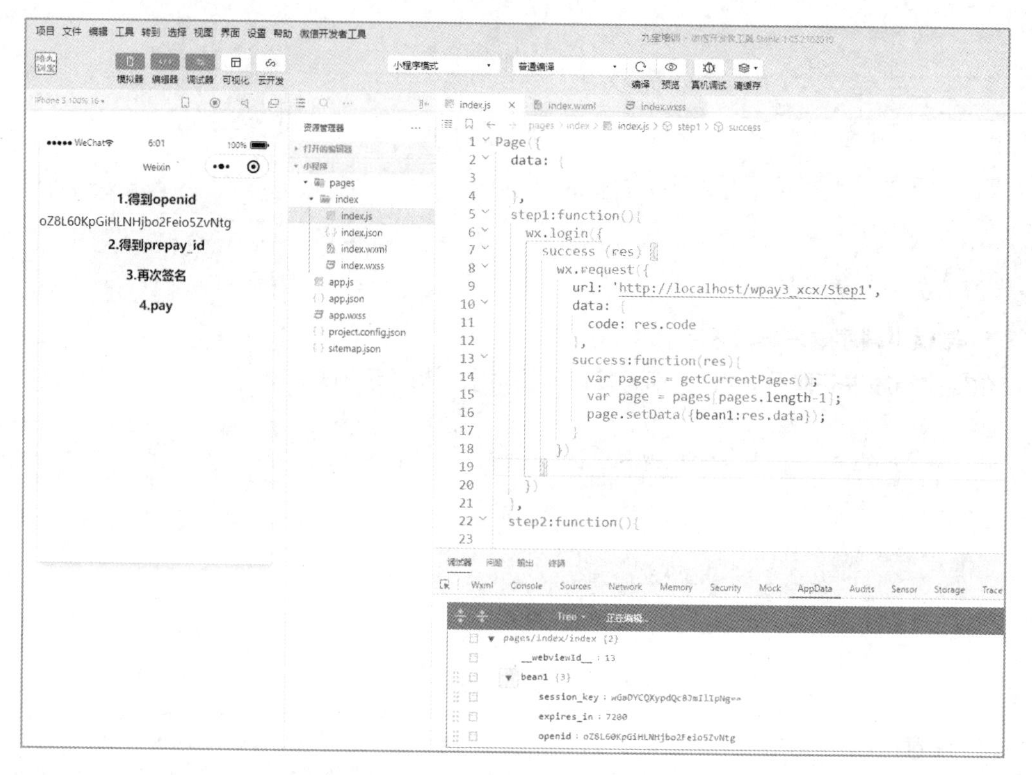

图 7-17

7.2.2 调用“统一下单 API”获取 prepay_id

“统一下单 API”为我们提供了微信支付接口，其调用计算方式以及参数加密算法较为复杂，因此这里直接给出实现程序，以供读者参考。对其中详细的算法感兴趣的读者可参考微信支付官方文档中的相关说明。

在服务端创建 servlet2，相关程序如下。

```
package jiubao2326321088;

import java.io.IOException;
import javax.servlet.ServletException;
import javax.servlet.annotation.WebServlet;
import javax.servlet.http.HttpServlet;
import javax.servlet.http.HttpServletRequest;
import javax.servlet.http.HttpServletResponse;

import org.apache.http.client.fluent.Request;
import org.apache.http.entity.ContentType;

import jiubao2326321088.util.WXUtil;
import jiubao2326321088.util.WX_Args;

@WebServlet("/Step2")
public class Step2 extends HttpServlet {
protected void doGet(HttpServletRequest request, HttpServletResponse response) throws
ServletException, IOException {
```

```
        String openid = request.getParameter("openid");
        String total  = "1";
        String out_trade_no = "jiubao_qq_2326321088_"+System.currentTimeMillis()/1000;

        String body = "{\"appid\":\""+WX_Args.appid
                +"\",\"mchid\":\""+WX_Args.mchid
                +"\",\"description\":\"Image 形象店-深圳腾大-QQ 公仔\""
                + ",\"out_trade_no\":\""+out_trade_no+"\""
                + ",\"notify_url\":\"https://www.weixin.qq.com/wxpay/pay.php\""
                + ",\"amount\":{\"total\":"+total+",\"currency\":\"CNY\"}"
                + ",\"payer\":{\"openid\":\""+openid+"\"}}";

        try {
            String Authorization = "WECHATPAY2-SHA256-RSA2048 " + WXUtil.getToken("POST",
"/v3/pay/transactions/jsapi", body);

            String str = "";
            try {
                str =
Request.Post("https://api.mch.weixin.qq.com/v3/pay/transactions/jsapi")
                        .addHeader("Accept", "application/json")
                        .addHeader("Content-Type", "application/json")
                        .addHeader("Authorization", Authorization)
                        .bodyString(body, ContentType.APPLICATION_JSON)
                        .execute()
                        .returnContent()
                        .asString();
            } catch (Exception e) {
                e.printStackTrace();
            }
            response.getWriter().print(str);
            System.out.println("Step2 " + str);
        } catch (Exception e1) {
            e1.printStackTrace();
        }
    }

    }
```

servlet2 的作用是先获取微信小程序端提交的 openid 参数，然后利用 openid 参数调用“统一下单 API”，响应本次请求。

WXUtil.getToken 的作用是获取签名相关信息，其中用到的相关方法如下。

```
    public static String getToken(String method, String url, String body) throws Exception
{
        long timestamp = System.currentTimeMillis() / 1000;
        String nonceStr = System.currentTimeMillis()+"";
        String message = buildMessage(method, url, timestamp, nonceStr, body);
        String signature = sign(message.getBytes("utf-8"));
        return "mchid=\"" + WX_Args.mchid + "\"," + "nonce_str=\"" + nonceStr + "\"," +
"timestamp=\"" + timestamp + "\"," + "serial_no=\"" + WX_Args.serial_no + "\"," +
"signature=\"" + signature + "\"";
    }
    public static String buildMessage(String method, String url, long timestamp, String
nonceStr, String body) {
        String str = method + "\n" + url + "\n" + timestamp + "\n" + nonceStr + "\n" +
body + "\n";
        return str;
    }
    public static String sign(byte[] message) throws Exception {
        PrivateKey yourPrivateKey = WXUtil.getPrivateKey(WX_Args.file);
        Signature sign = Signature.getInstance("SHA256withRSA");
        sign.initSign(yourPrivateKey);
```

```
        sign.update(message);
        return Base64.getEncoder().encodeToString(sign.sign());
    }
    public static PrivateKey getPrivateKey(String filename) throws Exception {
        String content = new String(Files.readAllBytes(Paths.get(filename)), "utf-8");
        try {
            String privateKey = content.replace("-----BEGIN PRIVATE KEY-----",
"").replace("-----END PRIVATE KEY-----", "").replaceAll("\\s+", "");
            KeyFactory kf = KeyFactory.getInstance("RSA");
            return kf.generatePrivate(new
PKCS8EncodedKeySpec(Base64.getDecoder().decode(privateKey)));
        } catch (NoSuchAlgorithmException e) {
            throw new RuntimeException("当前 Java 环境不支持 RSA", e);
        } catch (InvalidKeySpecException e) {
            throw new RuntimeException("无效的密钥格式");
        }
    }
```

微信小程序端 wx.request 请求 Web Server 携带 openid 参数，参考代码如下。

```
wx.request({
    url: 'http://localhost/wpay3_xcx/Step2',
    data:{
        openid:this.data.bean1.openid
    },
    success:function(res){
        var pages = getCurrentPages();
        var page = pages[pages.length-1];
        page.setData({bean2:res.data});
    }
})
```

与调用微信小程序 wx.request 获取 openid 相似，考虑到实际编程情况，这里使用 getCurrentPages()获取当前页面栈，进而得到当前 page。同样，对于 wx.request 请求失败的情况，读者可以酌情自行解决，如图 7-18 所示。

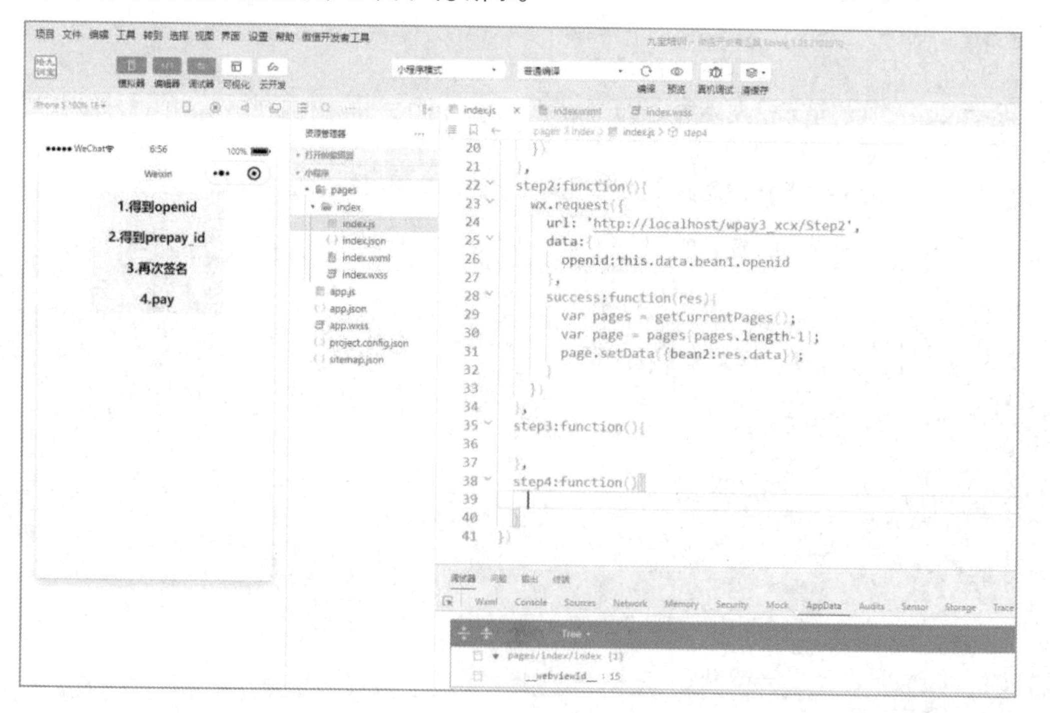

图 7-18

程序执行编译后，在模拟器中依次单击“1.得到 openid”和“2.得到 prepay_id”按钮，结果如图 7-19 所示。

图 7-19

7.2.3　再次签名

微信小程序调用支付功能时，需要相应参数进行服务端计算。

在服务端创建 servlet，参考代码如下。

```
package jiubao2326321088;

import java.io.IOException;
import javax.servlet.ServletException;
import javax.servlet.annotation.WebServlet;
import javax.servlet.http.HttpServlet;
import javax.servlet.http.HttpServletRequest;
import javax.servlet.http.HttpServletResponse;

import com.google.gson.Gson;

import jiubao2326321088.bean.BeanStep3;
import jiubao2326321088.util.WXUtil;

@WebServlet("/Step3")
public class Step3 extends HttpServlet {
    protected void doGet(HttpServletRequest request, HttpServletResponse response)
throws ServletException, IOException {
        String _timeStamp = System.currentTimeMillis()+"";
        String _nonceStr  = System.currentTimeMillis()+"";
        String prepay_id  = request.getParameter("prepay_id");
        BeanStep3 beanStep3 = new BeanStep3(_timeStamp, _nonceStr,
WXUtil.getPaySign(_timeStamp, _nonceStr, prepay_id));

        response.getWriter().print(new Gson().toJson(beanStep3));
```

```
            System.out.println("Step3   "+new Gson().toJson(beanStep3));
        }
    }
```

WXUtil.getPaySign 可以实现相关算法，参考代码如下。

```
    public static String getPaySign(String timestamp, String nonceStr, String prepayId)
{
        String str = WX_Args.appid + "\n"
                + timestamp + "\n"
                + nonceStr + "\n"
                + "prepay_id=" + prepayId + "\n";
        try {
            return sign(str.getBytes("utf-8"));
        } catch (Exception e) {
            e.printStackTrace();
            return null;
        }
    }
```

微信小程序端实现代码如下。

```
    wx.request({
        url: 'http://localhost/wpay3_xcx/Step3',
        data: {
            prepay_id: this.data.bean2.prepay_id,
        },
        success: function (res) {
            var pages = getCurrentPages();
            var page = pages[pages.length - 1];
            page.setData({ bean3: res.data });
        }
    })
```

程序执行编译后，在模拟器中依次单击“1.得到 openid”“2.得到 prepay_id”“3.再次签名”按钮，结果如图 7-20 所示。

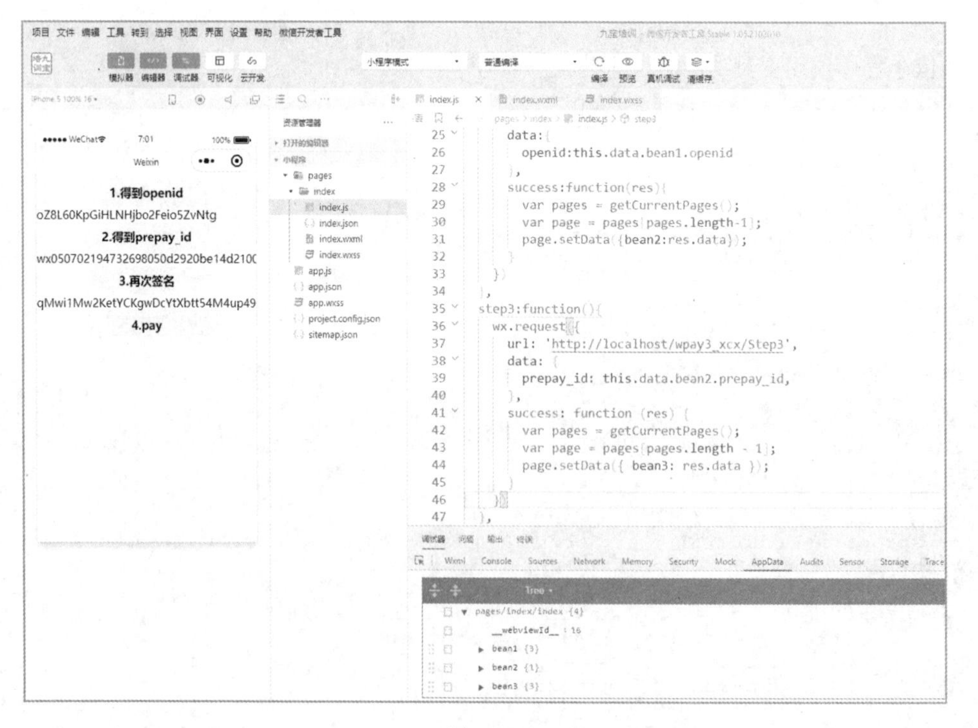

图 7-20

与调用微信小程序 wx.request 获取 openid 相似，考虑到实际开发情况，这里使用 getCurrentPages()获取当前页面栈，进而得到当前 page。同样，对于 wx.request 请求失败的情况，读者可以自行酌情解决。

7.2.4　调用微信支付功能

微信小程序端需要调用 wx.requestPayment 发起微信支付，参考代码如下。

```
wx.requestPayment({
     'timeStamp': this.data.bean3.timeStamp,
     'nonceStr': this.data.bean3.nonceStr,
     'package': 'prepay_id=' + this.data.bean2.prepay_id,
     'signType': 'RSA',
     'paySign': this.data.bean3.paySign,
     'success': function (res) {
         console.log(res);
     },
     'fail': function (res) {
         console.log(res);
     }
})
```

程序执行编译后，在模拟器中依次从上到下单击“1.得到 openid”“2.得到 prepay_id”“3.再次签名”“4.pay”按钮，结果如图 7-21 所示。

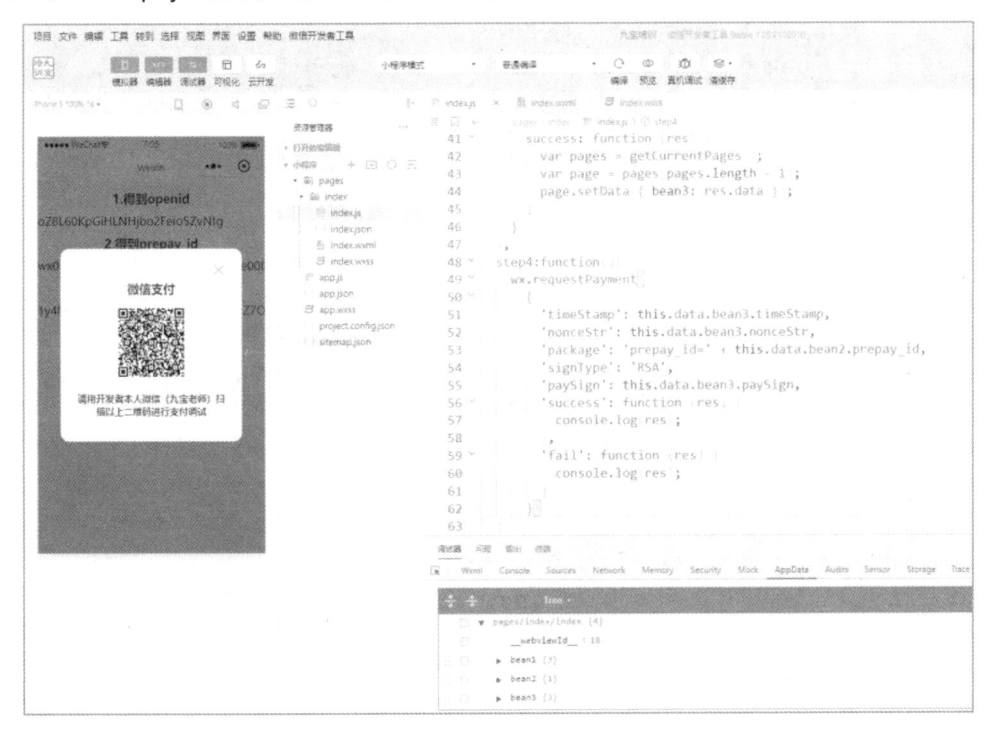

图 7-21

▶ **注意：**

开发工具显示的二维码是提供给手机测试使用的，并非展示给用户的。

7.3 程序清单

微信支付程序较为复杂，下面将全部程序以列表方式提供给读者，以供参考。

7.3.1 服务端

下面给出 7 组服务端代码。

（1）BeanStep1 代码如下。

```
package jiubao2326321088.bean;

public class BeanStep1 {

  private String openid;
  private String session_key;
  private String unionid;
  private String errcode;
  private String errmsg;

  public String toString() {
    return "BeanStep1 [openid=" + openid + ", session_key=" + session_key + ", unionid="
+ unionid + ", errcode=" + errcode + ", errmsg=" + errmsg + "]";
  }
  public String getOpenid() {
    return openid;
  }
  public void setOpenid(String openid) {
    this.openid = openid;
  }
  public String getSession_key() {
    return session_key;
  }
  public void setSession_key(String session_key) {
    this.session_key = session_key;
  }
  public String getUnionid() {
    return unionid;
  }
  public void setUnionid(String unionid) {
    this.unionid = unionid;
  }
  public String getErrcode() {
    return errcode;
  }
  public void setErrcode(String errcode) {
    this.errcode = errcode;
  }
  public String getErrmsg() {
    return errmsg;
  }
  public void setErrmsg(String errmsg) {
```

```
    this.errmsg = errmsg;
  }
}
```

（2）BeanStep3 代码如下。

```
package jiubao2326321088.bean;

public class BeanStep3 {

  private String timeStamp = "";
  private String nonceStr  = "";
  private String paySign   = "";

  public BeanStep3(String timeStamp, String nonceStr, String paySign) {
    super();
    this.timeStamp = timeStamp;
    this.nonceStr = nonceStr;
    this.paySign = paySign;
  }
  public String getTimeStamp() {
    return timeStamp;
  }
  public void setTimeStamp(String timeStamp) {
    this.timeStamp = timeStamp;
  }
  public String getNonceStr() {
    return nonceStr;
  }
  public void setNonceStr(String nonceStr) {
    this.nonceStr = nonceStr;
  }
  public String getPaySign() {
    return paySign;
  }
  public void setPaySign(String paySign) {
    this.paySign = paySign;
  }

}
```

（3）WX_Args 代码如下。

```
package jiubao2326321088.util;

public class WX_Args {

  public static String appid = "";
  public static String secret = "";
  public static String mchid = "";
  public static String key = "jiubao2326321088jiubao2326321088";
  public static String serial_no = "";
  public static String file = "E:\\九宝培训
\\1434507702_20210125_cert\\apiclient_key.pem";

}
```

（4）WXUtil 代码如下。

```
package jiubao2326321088.util;

import java.io.IOException;
import java.nio.file.Files;
import java.nio.file.Paths;
import java.security.KeyFactory;
import java.security.NoSuchAlgorithmException;
import java.security.PrivateKey;
import java.security.Signature;
import java.security.spec.InvalidKeySpecException;
import java.security.spec.PKCS8EncodedKeySpec;
import java.util.Base64;

import org.apache.http.client.ClientProtocolException;
import org.apache.http.client.fluent.Request;

import com.google.gson.Gson;

import jiubao2326321088.bean.BeanStep1;

public class WXUtil {

  public static String getopenid(String js_code){
    try {
      String str = Request.Get("https://api.weixin.qq.com/sns/jscode2session?appid="
+ WX_Args.appid + "&secret=" + WX_Args.secret + "&js_code=" + js_code +
"&grant_type=authorization_code")
              .execute()
              .returnContent()
              .asString();
      return str;
    } catch (Exception e) {
      e.printStackTrace();
      return null;
    }
  }

  public static String getToken(String method, String url, String body) throws Exception
{
    long timestamp = System.currentTimeMillis() / 1000;
    String nonceStr = System.currentTimeMillis()+"";
    String message = buildMessage(method, url, timestamp, nonceStr, body);
    String signature = sign(message.getBytes("utf-8"));
    return "mchid=\"" + WX_Args.mchid + "\"," + "nonce_str=\"" + nonceStr + "\"," +
"timestamp=\"" + timestamp + "\"," + "serial_no=\"" + WX_Args.serial_no + "\"," +
"signature=\"" + signature + "\"";
  }
  public static String buildMessage(String method, String url, long timestamp, String
nonceStr, String body) {
    String str = method + "\n" + url + "\n" + timestamp + "\n" + nonceStr + "\n" + body
+ "\n";
    return str;
  }
  public static String sign(byte[] message) throws Exception {
```

```
        PrivateKey yourPrivateKey = WXUtil.getPrivateKey(WX_Args.file);
        Signature sign = Signature.getInstance("SHA256withRSA");
        sign.initSign(yourPrivateKey);
        sign.update(message);
        return Base64.getEncoder().encodeToString(sign.sign());
      }
      public static PrivateKey getPrivateKey(String filename) throws Exception {
        String content = new String(Files.readAllBytes(Paths.get(filename)), "utf-8");
        try {
          String privateKey = content.replace("-----BEGIN PRIVATE KEY-----",
"").replace("-----END PRIVATE KEY-----", "").replaceAll("\\s+", "");
          KeyFactory kf = KeyFactory.getInstance("RSA");
          return kf.generatePrivate(new
PKCS8EncodedKeySpec(Base64.getDecoder().decode(privateKey)));
        } catch (NoSuchAlgorithmException e) {
          throw new RuntimeException("当前 Java 环境不支持 RSA", e);
        } catch (InvalidKeySpecException e) {
          throw new RuntimeException("无效的密钥格式");
        }
      }
      public static String getPaySign(String timestamp, String nonceStr, String prepayId)
{
        String str = WX_Args.appid + "\n"
                   + timestamp + "\n"
                   + nonceStr + "\n"
                   + "prepay_id=" + prepayId + "\n";
        try {
          return sign(str.getBytes("utf-8"));
        } catch (Exception e) {
          e.printStackTrace();
          return null;
        }
      }
    }
```

（5）Step1 代码如下。

```
    package jiubao2326321088;

    import java.io.IOException;
    import javax.servlet.ServletException;
    import javax.servlet.annotation.WebServlet;
    import javax.servlet.http.HttpServlet;
    import javax.servlet.http.HttpServletRequest;
    import javax.servlet.http.HttpServletResponse;

    import jiubao2326321088.bean.BeanStep1;
    import jiubao2326321088.util.WXUtil;

    @WebServlet("/Step1")
    public class Step1 extends HttpServlet {
      protected void doGet(HttpServletRequest request, HttpServletResponse response)
throws ServletException, IOException {
        String code = request.getParameter("code");
        System.out.println("step1");
```

```
        response.getWriter().print(WXUtil.getopenid(code));
      }
    }
```

（6） Step2 代码如下。

```
    package jiubao2326321088;

    import java.io.IOException;
    import javax.servlet.ServletException;
    import javax.servlet.annotation.WebServlet;
    import javax.servlet.http.HttpServlet;
    import javax.servlet.http.HttpServletRequest;
    import javax.servlet.http.HttpServletResponse;

    import org.apache.http.client.fluent.Request;
    import org.apache.http.entity.ContentType;

    import jiubao2326321088.util.WXUtil;
    import jiubao2326321088.util.WX_Args;

    @WebServlet("/Step2")
    public class Step2 extends HttpServlet {
      protected void doGet(HttpServletRequest request, HttpServletResponse response)
throws ServletException, IOException {

        String openid = request.getParameter("openid");
        String total  = "1";
        String out_trade_no = "jiubao_qq_2326321088_"+System.currentTimeMillis()/1000;

        String body = "{\"appid\":\""+WX_Args.appid
                +"\",\"mchid\":\""+WX_Args.mchid
                +"\",\"description\":\"Image 形象店-深圳腾大-QQ 公仔\""
                + ",\"out_trade_no\":\""+out_trade_no+"\""
                + ",\"notify_url\":\"https://www.weixin.qq.com/wxpay/pay.php\""
                + ",\"amount\":{\"total\":"+total+",\"currency\":\"CNY\"}"
                + ",\"payer\":{\"openid\":\""+openid+"\"}}";

        try {
          String Authorization = "WECHATPAY2-SHA256-RSA2048 " + WXUtil.getToken("POST",
"/v3/pay/transactions/jsapi", body);

          String str = "";
          try {
            str = Request.Post("https://api.mch.weixin.qq.com/v3/pay/transactions/jsapi")
                .addHeader("Accept", "application/json")
                .addHeader("Content-Type", "application/json")
                .addHeader("Authorization", Authorization)
                .bodyString(body, ContentType.APPLICATION_JSON)
                .execute()
                .returnContent()
                .asString();
          } catch (Exception e) {
            e.printStackTrace();
          }
          response.getWriter().print(str);
```

```
            System.out.println("Step2 " + str);
        } catch (Exception e1) {
            e1.printStackTrace();
        }
    }

}
```

（7）Step3 代码如下。

```
package jiubao2326321088;

import java.io.IOException;
import javax.servlet.ServletException;
import javax.servlet.annotation.WebServlet;
import javax.servlet.http.HttpServlet;
import javax.servlet.http.HttpServletRequest;
import javax.servlet.http.HttpServletResponse;

import com.google.gson.Gson;

import jiubao2326321088.bean.BeanStep3;
import jiubao2326321088.util.WXUtil;

@WebServlet("/Step3")
public class Step3 extends HttpServlet {
    protected void doGet(HttpServletRequest request, HttpServletResponse response)
throws ServletException, IOException {
        String _timeStamp = System.currentTimeMillis()+"";
        String _nonceStr  = System.currentTimeMillis()+"";
        String prepay_id  = request.getParameter("prepay_id");
        BeanStep3 beanStep3 = new BeanStep3(_timeStamp, _nonceStr,
WXUtil.getPaySign(_timeStamp, _nonceStr, prepay_id));

        response.getWriter().print(new Gson().toJson(beanStep3));
        System.out.println("Step3   "+new Gson().toJson(beanStep3));
    }

}
```

服务端用到的 JAR 文件共有 8 个，具体如下。

- commons-codec-1.9.jar。
- commons-logging-1.2.jar。
- fluent-hc-4.5.3.jar。
- gson.jar。
- httpclient-4.5.3.jar。
- httpcore-4.4.6.jar。
- slf4j-api-1.6.6.jar。
- slf4j-log4j12-1.6.6.jar。

付款通知与订单查询相关内容，感兴趣的读者参见笔者的《移动支付开发实战》一书。

7.3.2 小程序端

小程序端代码包括 JS 和 WXML 两部分。

（1）JS 文件代码如下。

```
Page({
  data: {

  },
  step1:function(){
    wx.login({
      success (res) {
        wx.request({
          url: 'http://localhost/wpay3_xcx/Step1',
          data: {
            code: res.code
          },
          success:function(res){
            var pages = getCurrentPages();
            var page = pages[pages.length-1];
            page.setData({bean1:res.data});
          }
        })
      }
    })
  },
  step2:function(){
    wx.request({
      url: 'http://localhost/wpay3_xcx/Step2',
      data:{
        openid:this.data.bean1.openid
      },
      success:function(res){
        var pages = getCurrentPages();
        var page = pages[pages.length-1];
        page.setData({bean2:res.data});
      }
    })
  },
  step3:function(){
    wx.request({
      url: 'http://localhost/wpay3_xcx/Step3',
      data: {
        prepay_id: this.data.bean2.prepay_id,
      },
      success: function (res) {
        var pages = getCurrentPages();
        var page = pages[pages.length - 1];
        page.setData({ bean3: res.data });
      }
    })
  },
  step4:function(){
    wx.requestPayment(
```

```
      {
        'timeStamp': this.data.bean3.timeStamp,
        'nonceStr': this.data.bean3.nonceStr,
        'package': 'prepay_id=' + this.data.bean2.prepay_id,
        'signType': 'RSA',
        'paySign': this.data.bean3.paySign,
        'success': function (res) {
          console.log(res);
        },
        'fail': function (res) {
          console.log(res);
        }
      })
  }
})
```

（2）WXML 代码如下。

```
<button bindtap="step1">1.得到 openid</button>
<view>{{bean1.openid}}</view>
<button bindtap="step2">2.得到 prepay_id</button>
<view>{{bean2.prepay_id}}</view>
<button bindtap="step3">3.再次签名</button>
<view>{{bean3.paySign}}</view>
<button bindtap="step4">4.pay</button>
```

第 8 章　小程序商城

微信小程序由于其自身的特殊性，开发方式与传统应用程序开发有着很大的区别。本章通过一个微信小程序商城案例，讲解小程序项目开发的相关知识。

8.1　项目概述

本案例微信小程序商城主要包含“商品列表展示”“商品详情展示”“购物车”3 个功能页面。

在“商品列表展示”页面可访问微信小程序的后台服务，将数据库中记录的商品信息以特定报文格式传送到微信小程序前端。同时，小程序按照商品类别分为百货、餐厨、办公、游戏、动漫 5 类，提供商品图片、商品名称、价格、包邮、活动商品展示等信息，如图 8-1 所示。

在“商品详情展示”页面主要提供商品的详情图片、标签、名称、简介、包邮、购物券、全国联保、运费险、买家留言等信息，如图 8-2 所示。用户可以随意修改商品购买数量，修改后，系统会根据购买数量和商品单价自动计算单件商品的总价，保存数据并在购物车中展示。

在“购物车”页面可展示拟购买商品的名称、数量、单价，自动核算订单总价，并提醒用户填写收件人、邮递地址和联系电话信息。用户填写完收件信息后，单击 submit 按钮，订单信息将提交给后端服务器；如果收件信息填写错误，可单击 reset 按钮，系统将清空用户已填写的收件人、邮递地址、联系电话信息，用户可再次输入，如图 8-3 所示。

图 8-1

图 8-2

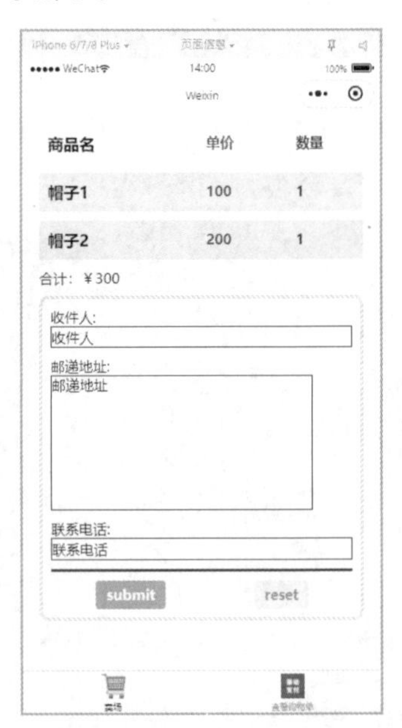

图 8-3

8.2　数据库设计

开发本项目时，一共要用到 5 张数据表，分别是 sequ 表、shangpin 表、shangpinlog 表、userinfo 表和 xiaoshou 表。

- sequ 表：记录订单索引数。
- shangpin 表：记录商品信息。
- shangpinlog 表：记录对应商品信息下的买家留言。
- userinfo 表：记录订单联系人。
- xiaoshou 表：记录订单详情。

建表相关的 SQL 代码如下。

```
/*
 Navicat MySQL Data Transfer

 Source Server         : localhost
 Source Server Type    : MySQL
 Source Server Version : 80019
 Source Host           : localhost:3306
 Source Schema         : shop

 Target Server Type    : MySQL
 Target Server Version : 80019
 File Encoding         : 65001

 Date: 05/04/2021 18:47:09
*/

SET NAMES utf8mb4;
SET FOREIGN_KEY_CHECKS = 0;

-- ----------------------------
-- Table structure for sequ
-- ----------------------------
DROP TABLE IF EXISTS `sequ`;
CREATE TABLE `sequ`  (
  `id` int(0) NOT NULL AUTO_INCREMENT,
  `col_name` varchar(50) CHARACTER SET utf8 COLLATE utf8_general_ci NOT NULL DEFAULT
'0' COMMENT '索引名',
  `code` varchar(50) CHARACTER SET utf8 COLLATE utf8_general_ci NOT NULL DEFAULT '0'
COMMENT '索引值',
  `text` varchar(50) CHARACTER SET utf8 COLLATE utf8_general_ci NULL DEFAULT NULL
COMMENT '说明',
  PRIMARY KEY (`id`) USING BTREE,
  UNIQUE INDEX `col_name`(`col_name`) USING BTREE
) ENGINE = InnoDB AUTO_INCREMENT = 1 CHARACTER SET = utf8 COLLATE = utf8_general_ci
ROW_FORMAT = Compact;

-- ----------------------------
-- Table structure for shangpin
-- ----------------------------
DROP TABLE IF EXISTS `shangpin`;
CREATE TABLE `shangpin`  (
  `id` int(0) UNSIGNED NOT NULL AUTO_INCREMENT,
  `typecode` varchar(255) CHARACTER SET utf8 COLLATE utf8_general_ci NULL DEFAULT NULL
```

```
COMMENT '类型编号',
      `typename` varchar(255) CHARACTER SET utf8 COLLATE utf8_general_ci NULL DEFAULT NULL
COMMENT '类型名',
      `shangpincode` varchar(255) CHARACTER SET utf8 COLLATE utf8_general_ci NULL DEFAULT
NULL COMMENT '商品编号',
      `shangpinname` varchar(255) CHARACTER SET utf8 COLLATE utf8_general_ci NULL DEFAULT
NULL COMMENT '商品名',
      `logo` varchar(255) CHARACTER SET utf8 COLLATE utf8_general_ci NULL DEFAULT '' COMMENT
'标签',
      `price` decimal(10, 2) UNSIGNED NOT NULL COMMENT '价格',
      `img` varchar(255) CHARACTER SET utf8 COLLATE utf8_general_ci NULL DEFAULT NULL
COMMENT '图片',
      `baoyou` varchar(255) CHARACTER SET utf8 COLLATE utf8_general_ci NULL DEFAULT ''
COMMENT '包邮',
      `gouwuquan` varchar(255) CHARACTER SET utf8 COLLATE utf8_general_ci NULL DEFAULT ''
COMMENT '购物券',
      `zengpinbaozheng` varchar(255) CHARACTER SET utf8 COLLATE utf8_general_ci NULL
DEFAULT '' COMMENT '正品保证',
      `quanguolianbao` varchar(255) CHARACTER SET utf8 COLLATE utf8_general_ci NULL DEFAULT
'' COMMENT '全国联保',
      `zengyunfeixian` varchar(255) CHARACTER SET utf8 COLLATE utf8_general_ci NULL DEFAULT
'' COMMENT '赠运费险',
      `text` varchar(1000) CHARACTER SET utf8 COLLATE utf8_general_ci NOT NULL DEFAULT ''
COMMENT '简介',
      `odb` int(0) NOT NULL DEFAULT 0,
      `createdate` timestamp(0) NOT NULL DEFAULT CURRENT_TIMESTAMP(0) ON UPDATE
CURRENT_TIMESTAMP(0),
      PRIMARY KEY (`id`) USING BTREE
    ) ENGINE = InnoDB AUTO_INCREMENT = 1 CHARACTER SET = utf8 COLLATE = utf8_general_ci
COMMENT = '菜谱' ROW_FORMAT = Dynamic;

    -- ----------------------------
    -- Table structure for shangpinlog
    -- ----------------------------
    DROP TABLE IF EXISTS `shangpinlog`;
    CREATE TABLE `shangpinlog`  (
      `id` int(0) UNSIGNED NOT NULL AUTO_INCREMENT,
      `typecode` varchar(255) CHARACTER SET utf8 COLLATE utf8_general_ci NULL DEFAULT NULL
COMMENT '类别编号',
      `typename` varchar(255) CHARACTER SET utf8 COLLATE utf8_general_ci NULL DEFAULT NULL
COMMENT '类别名',
      `shangpincode` varchar(255) CHARACTER SET utf8 COLLATE utf8_general_ci NULL DEFAULT
NULL COMMENT '商品编号',
      `shangpinname` varchar(255) CHARACTER SET utf8 COLLATE utf8_general_ci NULL DEFAULT
NULL COMMENT '商品名',
      `text` varchar(1000) CHARACTER SET utf8 COLLATE utf8_general_ci NOT NULL DEFAULT ''
COMMENT '菜品简介',
      `nick_name` varchar(255) CHARACTER SET utf8 COLLATE utf8_general_ci NULL DEFAULT NULL
COMMENT '昵称',
      `avatar_url` varchar(255) CHARACTER SET utf8 COLLATE utf8_general_ci NULL DEFAULT
NULL COMMENT '头像',
      `createdate` timestamp(0) NOT NULL DEFAULT CURRENT_TIMESTAMP(0) ON UPDATE
CURRENT_TIMESTAMP(0),
      PRIMARY KEY (`id`) USING BTREE
    ) ENGINE = InnoDB AUTO_INCREMENT = 1 CHARACTER SET = utf8 COLLATE = utf8_general_ci
ROW_FORMAT = Dynamic;

    -- ----------------------------
    -- Table structure for userinfo
    -- ----------------------------
    DROP TABLE IF EXISTS `userinfo`;
    CREATE TABLE `userinfo`  (
```

```
    `id` int(0) UNSIGNED NOT NULL AUTO_INCREMENT,
    `orderid` varchar(255) CHARACTER SET utf8 COLLATE utf8_general_ci NOT NULL COMMENT
'订单 id',
    `nick_name` varchar(255) CHARACTER SET utf8 COLLATE utf8_general_ci NULL DEFAULT NULL
COMMENT '昵称',
    `avatar_url` varchar(255) CHARACTER SET utf8 COLLATE utf8_general_ci NULL DEFAULT
NULL COMMENT '头像',
    `gender` varchar(255) CHARACTER SET utf8 COLLATE utf8_general_ci NULL DEFAULT NULL
COMMENT '性别',
    `province` varchar(255) CHARACTER SET utf8 COLLATE utf8_general_ci NULL DEFAULT NULL
COMMENT '省份\n\n',
    `city` varchar(255) CHARACTER SET utf8 COLLATE utf8_general_ci NULL DEFAULT NULL
COMMENT '城市',
    `sjr` varchar(255) CHARACTER SET utf8 COLLATE utf8_general_ci NULL DEFAULT NULL
COMMENT '收件人',
    `dz` varchar(255) CHARACTER SET utf8 COLLATE utf8_general_ci NULL DEFAULT NULL COMMENT
'地址',
    `dh` varchar(255) CHARACTER SET utf8 COLLATE utf8_general_ci NULL DEFAULT NULL COMMENT
'电话',
    `createtime` timestamp(0) NULL DEFAULT NULL ON UPDATE CURRENT_TIMESTAMP(0),
    PRIMARY KEY (`id`) USING BTREE
  ) ENGINE = InnoDB AUTO_INCREMENT = 1 CHARACTER SET = utf8 COLLATE = utf8_general_ci
ROW_FORMAT = Dynamic;

  -- ----------------------------
  -- Table structure for xiaoshou
  -- ----------------------------
  DROP TABLE IF EXISTS `xiaoshou`;
  CREATE TABLE `xiaoshou`  (
    `id` int(0) UNSIGNED NOT NULL AUTO_INCREMENT,
    `orderid` varchar(255) CHARACTER SET utf8 COLLATE utf8_general_ci NULL DEFAULT NULL
COMMENT '订单 id',
    `count` int(0) NULL DEFAULT NULL COMMENT '数量',
    `typecode` varchar(255) CHARACTER SET utf8 COLLATE utf8_general_ci NULL DEFAULT NULL
COMMENT '类别编号',
    `typename` varchar(255) CHARACTER SET utf8 COLLATE utf8_general_ci NULL DEFAULT NULL
COMMENT '类别名',
    `shangpincode` varchar(255) CHARACTER SET utf8 COLLATE utf8_general_ci NULL DEFAULT
NULL COMMENT '商品编号',
    `shangpinname` varchar(255) CHARACTER SET utf8 COLLATE utf8_general_ci NULL DEFAULT
NULL COMMENT '商品名',
    `price` decimal(10, 2) NULL DEFAULT NULL COMMENT '价格',
    `img` varchar(255) CHARACTER SET utf8 COLLATE utf8_general_ci NULL DEFAULT NULL
COMMENT '图片',
    `text` varchar(255) CHARACTER SET utf8 COLLATE utf8_general_ci NULL DEFAULT NULL
COMMENT '说明',
    `createtime` timestamp(0) NULL DEFAULT NULL ON UPDATE CURRENT_TIMESTAMP(0),
    PRIMARY KEY (`id`) USING BTREE
  ) ENGINE = InnoDB AUTO_INCREMENT = 1 CHARACTER SET = utf8 COLLATE = utf8_general_ci
ROW_FORMAT = Dynamic;

  SET FOREIGN_KEY_CHECKS = 1;
```

8.3 “商品列表展示”页面

新建微信小程序项目，构建项目的“最小程序状态”。

由于项目需要访问服务端，开发阶段需单击微信小程序“详情”按钮，在弹出面板的“本地设置”选项卡下选中“不校验合法域名、web-view（业务域名）、TLS 版本以及 HTTPS 证书”复选框，如图 8-4 所示。

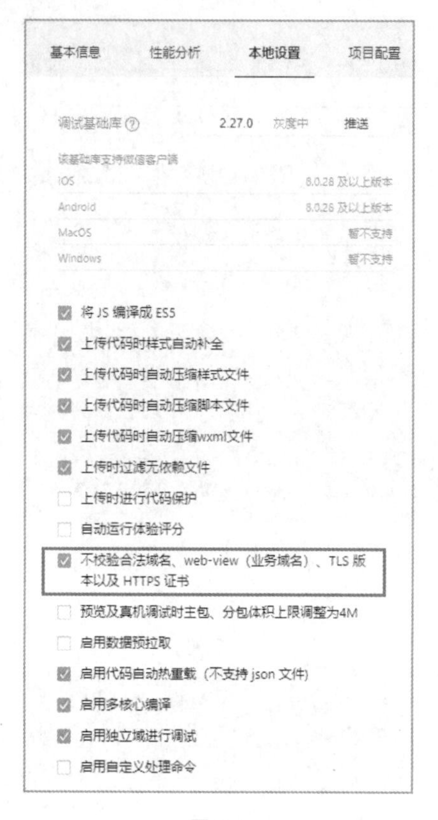

图 8-4

删除 index 文件，新建 shop 文件夹，新建 shop 页面，完成 app.json 配置，如图 8-5 所示。

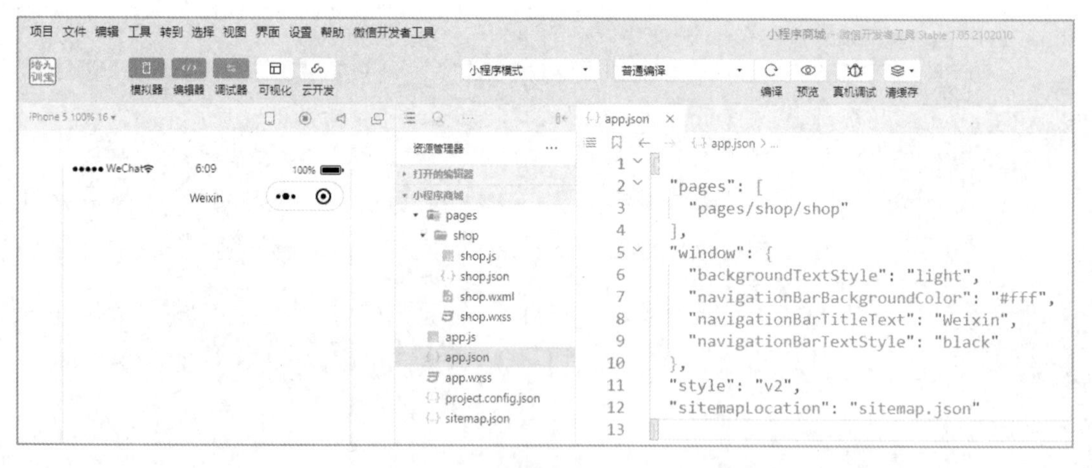

图 8-5

下面来实现小程序主页面顶部的商品分类导航菜单。分类导航功能可以使用微信小程序的原生组件来实现，本案例使用 button 组件来实现，效果如图 8-6 所示。

修改 shop.wxml 文件，增加以下程序代码。

```
    <view class='title_view'>
      <button class='title' type='primary' size='mini' disabled='{{10==typeCode}}' data-arg='10' bindtap="selectTitle">百货</button>
      <button class='title' type='primary' size='mini' disabled='{{20==typeCode}}' data-arg='20' bindtap="selectTitle">餐厨</button>
      <button class='title' type='primary' size='mini' disabled='{{30==typeCode}}' data-arg='30' bindtap="selectTitle">办公</button>
      <button class='title' type='primary' size='mini' disabled='{{40==typeCode}}' data-arg='40' bindtap="selectTitle">游戏</button>
      <button class='title' type='primary' size='mini' disabled='{{50==typeCode}}' data-arg='50' bindtap="selectTitle">动漫</button>
    </view>
```

修改 shop.js 文件，定义 button 事件，程序代码如下。

```
    selectTitle: function (event) {

        /**修改 title */
        this.setData({ typeCode: event.target.dataset.arg });

        this.showShangpin();

      },
```

用户单击按钮时，selectTitle 将响应事件，通过 event.target.dataset.arg 获取当前被单击的 button 参数，并将该参数同步给变量 typeCode。

shop.wxml 使用 disabled='{{10==typeCode}}'控制当前所选的 button 组件。这里需要定义变量 typeCode，初始值为 10。

此时的商品分类效果如图 8-7 所示。

图 8-6

图 8-7

商品列表展示需要先请求服务器，然后服务器请求数据库，构造报文返回微信小程序。定义微信小程序函数 showShangpin，参考代码如下。

```
showShangpin: function () {

  util.requestData(
    this.data.url + 'Shop',
    {
      type: 'showShangPin',
      typeCode: this.data.typeCode
    },
    function (res) {
      //得到服务器端数据
      var pages = getCurrentPages();
      var page = pages[pages.length - 1];
      page.setData({ shangpin: res.data });
    }
  );
},
```

微信小程序访问服务器封装的函数 util.js，代码如下。

```
var util = require('../../utils/util.js');
```

util.js 的实现代码如下。

```
function formatTime(date) {
  var year = date.getFullYear()
  var month = date.getMonth() + 1
  var day = date.getDate()

  var hour = date.getHours()
  var minute = date.getMinutes()
  var second = date.getSeconds()

  return [year, month, day].map(formatNumber).join('/') + ' ' + [hour, minute, sec
ond].map(formatNumber).join(':')
}

function formatNumber(n) {
  n = n.toString()
  return n[1] ? n : '0' + n
}

module.exports = {
  formatTime: formatTime,
  requestData : requestData
}

function requestData(url,data,success){
  wx.request({
    url: url,
    data: data,
    header: {
      'Content-Type': 'application/json'
    },
```

```
        method: 'get',
        success: success
    })
}
```

以上代码中使用了 3 个函数，其功能及作用如下。

◆ function formatTime(date)：格式化时间。

◆ function formatNumber(n)：格式化数字。

◆ function requestData(url,data,success)：调用服务器服务。

以下程序将 formatTime 与 requestData 对外暴露。

```
module.exports = {
  formatTime: formatTime,
  requestData : requestData
}
```

注意：

封装请求的意义是同意定义服务域名。此外，微服务架构的系统需要定义 Authentication。

这里按照最基础的请求响应方式予以讲解，以降低读者的学习门槛。

在服务端定义 servlet，程序代码如下。

```
package ajax;

import java.io.IOException;
import java.io.PrintWriter;
import java.text.SimpleDateFormat;
import java.util.Date;
import java.util.List;
import javax.servlet.ServletException;
import javax.servlet.annotation.WebServlet;
import javax.servlet.http.HttpServlet;
import javax.servlet.http.HttpServletRequest;
import javax.servlet.http.HttpServletResponse;
import com.google.gson.Gson;
import com.google.gson.GsonBuilder;
import com.google.gson.reflect.TypeToken;
import bean.ShangpinXiangqing;
import generator.model.Shangpin;
import generator.model.Shangpinlog;
import generator.model.Userinfo;
import generator.model.Xiaoshou;
import server.ShopService;
import util.WX_Util;

@WebServlet("/Shop")
public class Shop extends HttpServlet {

    protected void doGet(HttpServletRequest request, HttpServletResponse response)
throws ServletException, IOException {
        // TODO Auto-generated method stub
        doPost(request,response);
    }
```

```
    protected void doPost(HttpServletRequest request, HttpServletResponse response)
throws ServletException, IOException {

        request.setCharacterEncoding("UTF-8");
        response.setCharacterEncoding("UTF-8");

        PrintWriter out = response.getWriter();

        Gson gson = new GsonBuilder().setDateFormat("yyyy-MM-dd").create();

        String type= request.getParameter("type");
        System.out.println("type    "+type);

        if("showShangPin".equals(type)){

            Shangpin shangpin = new Shangpin();
            shangpin.setTypecode(request.getParameter("typeCode"));
            List<Shangpin> caipus = ShopService.queryShangpin(shangpin);
            out.print(gson.toJson(caipus));

        }else if("shopInfo".equals(type)){

            Shangpin shangpin = new Shangpin();
            shangpin.setShangpincode(request.getParameter("shangpincode"));

            List<Shangpin> caipus = ShopService.queryShangpin(shangpin);

            Shangpinlog shangpinlog = new Shangpinlog();
            shangpinlog.setShangpincode(request.getParameter("shangpincode"));

            List<Shangpinlog> shangpinlogs =
ShopService.queryShangpinlog(shangpinlog);

            ShangpinXiangqing shangpinXiangqing = new
ShangpinXiangqing(caipus.get(0),shangpinlogs);

            out.print(gson.toJson(shangpinXiangqing));

        }else if("saveOrder".equals(type)){

            String orderId = WX_Util.getSequ("OrderId");

            Date date = new Date();
            SimpleDateFormat simpleDateFormat = new
SimpleDateFormat("yyyyMMddHHmmss_");
            orderId = simpleDateFormat.format(date)+orderId;

            boolean flag = ShopService.saveOrder(
                    orderId,
                    date,
                    gson.fromJson(request.getParameter("userInfo"), Userinfo.class),
                    gson.fromJson(request.getParameter("userInfo_"),
Userinfo.class),
                    gson.fromJson(request.getParameter("YigouShangpin"), new
TypeToken<List<Xiaoshou>>(){}.getType()));
            if(flag){
```

```
                out.print("{\"flag\":\""+flag+"\",\"orderId\":\""+orderId+"\"}");
            }else{
                out.print("{\"flag\":\""+flag+"\"}");
            }

        }else if("".equals(type)){

        }else{
            out.println("type="+type);
        }

        out.flush();
        out.close();
    }

}
```

▶ **注意：**

（1）servlet 提供对外请求，本案例作为教学案例，意在说明小程序项目开发基础架构。对于正式项目，需要实现请求认证。

（2）servlet 通过判断 type 的值，确定应该使用哪个响应逻辑。

针对本次请求，代码如下。

```
util.requestData(
  this.data.url + 'Shop',
  {
    type: 'showShangPin',
    typeCode: this.data.typeCode
  },
  function (res) {
    //得到服务器端数据
    var pages = getCurrentPages();
    var page = pages[pages.length - 1];
    page.setData({ shangpin: res.data });
  }
);
```

代码“type: 'showShangPin'”说明请求的是商品列表数据。所以，服务器的程序是请求数据库，将请求到的数据格式化后返回微信小程序。

```
Shangpin shangpin = new Shangpin();
shangpin.setTypecode(request.getParameter("typeCode"));
List<Shangpin> caipus = ShopService.queryShangpin(shangpin);
out.print(gson.toJson(caipus));
```

ShopService 类实现对 Shangpin 表的数据查询、增加、删除、修改等，代码如下。

```
package server;

import util.DBUtil;

import java.util.Date;
import java.util.List;
```

```
    import org.apache.ibatis.session.SqlSession;
    import generator.client.ShangpinMapper;
    import generator.client.ShangpinlogMapper;
    import generator.client.UserinfoMapper;
    import generator.client.XiaoshouMapper;
    import generator.model.Shangpin;
    import generator.model.ShangpinExample;
    import generator.model.ShangpinExample.Criteria;
    import generator.model.Shangpinlog;
    import generator.model.ShangpinlogExample;
    import generator.model.Userinfo;
    import generator.model.Xiaoshou;

    public class ShopService {

    public static List<Shangpin> queryShangpin(Shangpin shangpin){

         SqlSession sqlSession = DBUtil.getSqlSession();
         ShangpinMapper shangpinMapper = sqlSession.getMapper(ShangpinMapper.class);

         ShangpinExample example = new ShangpinExample();
         Criteria criteria = example.createCriteria();

    if(null!=shangpin.getShangpincode()){criteria.andShangpincodeEqualTo(shangpin.get
Shangpincode());}

    if(null!=shangpin.getTypecode()){criteria.andTypecodeEqualTo(shangpin.getTypecode
());}
         List<Shangpin> ShopList = shangpinMapper.selectByExample(example );

         sqlSession.close();

         return ShopList;

    }

    public static List<Shangpinlog> queryShangpinlog(Shangpinlog shangpinlog){

         SqlSession sqlSession = DBUtil.getSqlSession();
         ShangpinlogMapper shangpinlogMapper =
sqlSession.getMapper(ShangpinlogMapper.class);
         ShangpinlogExample example = new ShangpinlogExample();
         ShangpinlogExample.Criteria criteria = example.createCriteria();

    if(null!=shangpinlog.getShangpincode()){criteria.andShangpincodeEqualTo(shangpinl
og.getShangpincode());}

    if(null!=shangpinlog.getTypecode()){criteria.andTypecodeEqualTo(shangpinlog.getTy
pecode());}
         List<Shangpinlog> ShopList = shangpinlogMapper.selectByExample(example );

         sqlSession.close();

         return ShopList;

    }
```

```
    public static boolean saveOrder(String orderId, Date date, Userinfo userInfo,Userinfo
userInfo_, List<Xiaoshou> xiaoshous){

        boolean flag = false;

        SqlSession sqlSession = DBUtil.getSqlSession();

        try {
            UserinfoMapper userinfoMapper =
sqlSession.getMapper(UserinfoMapper.class);
            userInfo.setOrderid(orderId);
            userInfo.setCreatetime(date);
            userInfo.setDh(userInfo_.getDh());
            userInfo.setDz(userInfo_.getDz());
            userInfo.setSjr(userInfo_.getSjr());
            userinfoMapper.insertSelective(userInfo);

            XiaoshouMapper xiaoshouMapper =
sqlSession.getMapper(XiaoshouMapper.class);

            for(int x = 0 ; x < xiaoshous.size() ; x++){
                xiaoshous.get(x).setId(null);
                xiaoshous.get(x).setOrderid(orderId);
                xiaoshous.get(x).setCreatetime(date);
                xiaoshouMapper.insertSelective(xiaoshous.get(x));
            }

            sqlSession.commit();
            flag = true;
        } catch (Exception e) {
            sqlSession.rollback();
            e.printStackTrace();
        }

        sqlSession.close();

        return flag;
    }

    }
```

本案例程序中用到了 generator 插件和 mybatis 框架。

对于当前页面，需要在加载时执行以下程序。

```
wx.setStorageSync('YigouShangpin', new Array());
this.setData({ url: getApp().url });
this.showShangpin();
```

◆　“wx.setStorageSync('YigouShangpin', new Array());”的作用是修改缓存变量。

◆　“this.setData({ url: getApp().url });”的作用是定义请求 url。

◆　“this.showShangpin();”的作用是执行请求。

app.js 定义请求的地址如下。

```
url: 'http://localhost/shop/'
```

启动服务器，编译微信小程序，服务器执行的 SQL 代码如下。

```
    select id, typecode, typename, shangpincode, shangpinname, logo, price, img, baoyou, gouwuquan, zengpinbaozheng, quanguolianbao, zengyunfeixian, text, odb, createdate from shangpin WHERE ( typecode = ? )
    DEBUG [http-nio-80-exec-2] - ==> Parameters: 10(String)
```

查询到符合条件的有 16 条记录。至此，已完成数据列表的数据请求。

使用微信小程序开发工具可以查阅小程序本地记录的数据，如图 8-8 所示。

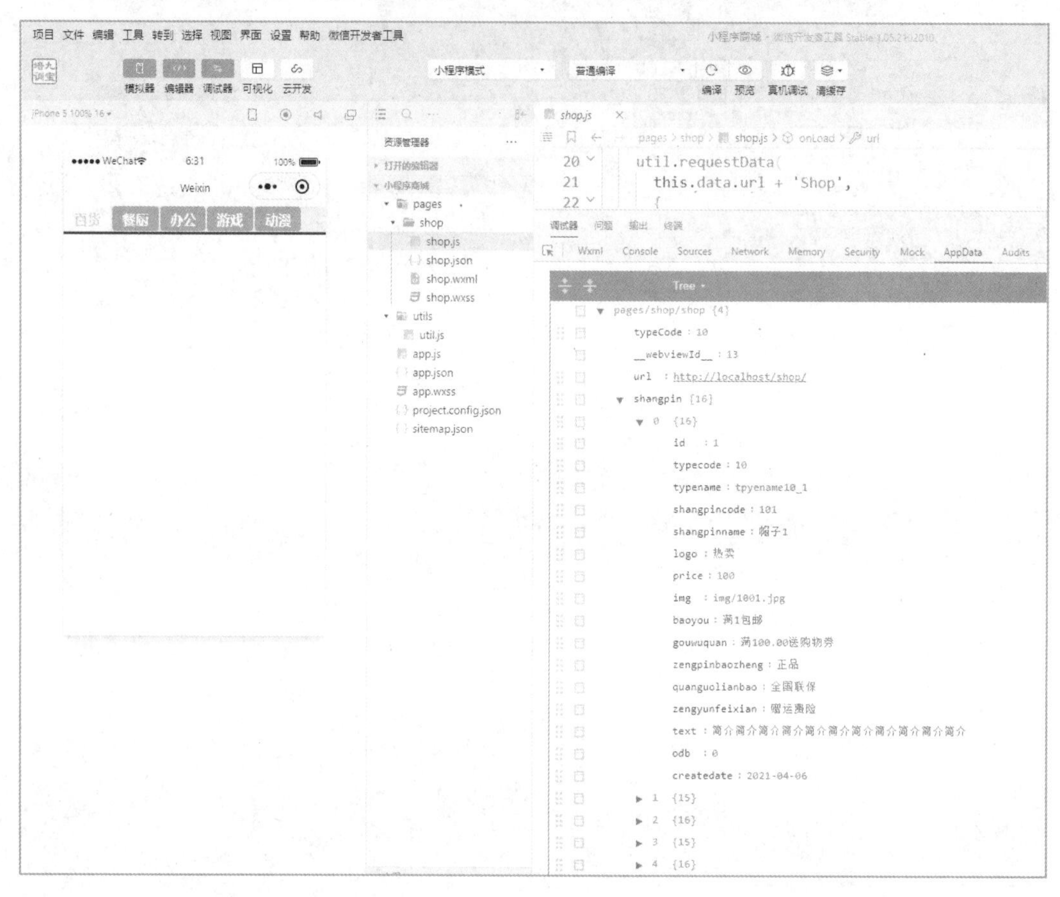

图 8-8

修改 shop.wxml 文件，程序代码如下。

```
    <view class='title_view'>
      <button class='title' type='primary' size='mini' disabled='{{10==typeCode}}' data-arg='10' bindtap="selectTitle">百货</button>
      <button class='title' type='primary' size='mini' disabled='{{20==typeCode}}' data-arg='20' bindtap="selectTitle">餐厨</button>
      <button class='title' type='primary' size='mini' disabled='{{30==typeCode}}' data-arg='30' bindtap="selectTitle">办公</button>
      <button class='title' type='primary' size='mini' disabled='{{40==typeCode}}' data-arg='40' bindtap="selectTitle">游戏</button>
      <button class='title' type='primary' size='mini' disabled='{{50==typeCode}}' data-arg='50' bindtap="selectTitle">动漫</button>
    </view>
```

```
    <scroll-view scroll-y='true' style='height:{{app.windowHeight-30}}px'>
      <view wx:for="{{shangpin}}" class='content' data-shangpincode='{{item.shangpinco
de}}' bindtap='choose'>
        <image class='shangpin_img' mode="aspectFit" src="{{url}}{{item.img}}" />
        <view class='context_text' style='width:{{app.windowWidth-150}}px'>
          <view style="height:50px">
            <text class='logo' wx:if='{{""!=item.logo}}'>{{item.logo}}</text>
            <text class="text">{{item.shangpinname}}</text>
          </view>
          <view style="height:20px;font-size:small">
            <text class='biaoqian' wx:if='{{""!=item.baoyou}}'>{{item.baoyou}}</text>
            <text class='biaoqian' wx:if='{{""!=item.gouwuquan}}'>{{item.gouwuquan}}</
text>
          </view>
          <view class="price">
            ￥: {{item.price}}
          </view>
        </view>
      </view>
    </scroll-view>
```

修改 shop.wxss 文件，程序代码如下。

```
    .title_view{
      border-bottom-style:solid;
      border-bottom-color:blue;
      width: 100%;height:30px;
      background-color: #eee;
    }

    .title{
      margin-left: 1px;
      margin-right: 1px;
      height:29px;
    }

    .content{
      display: flex;
      margin: 3px 1px 3px 1px;
      border-bottom: 1px dashed black;
    }

    .shangpin_img{
      width: 100px;
      height: 100px;
    }

    .context_text{
      padding: 5px
    }

    .logo{
      background-color:red;
      color:yellow;
      font-size:small;
      margin-top: 1px;
      width: 2ex;
```

```
}

.text{
  font-size:small
}

.biaoqian{
  color:red;
  border-color:red;
  border-style:solid;
  border-width: 1px;
  border-radius: 5px;
  margin: 1px;
  padding: 1px;
  font-size:small;
}

.price{
  color:red;
  font-size:small;
  height:20px;
  bottom:0;
}
```

完成的“商品列表展示”页面如图 8-9 所示。

图 8-9

调试器中关于微信小程序的提示信息如图 8-10 所示。

```
调试器　问题　输出　终端
Wxml　Console　Sources　Network　Memory　Security　Mock　AppData　Audits　Sensor　Storage　Trace
appservice　Filter　Default levels
[sitemap 索引情况提示] 根据 sitemap 的规则[0]，当前页面 [pages/shop/shop] 将被索引
▶WXMLRT_$gwx:./pages/shop/shop.wxml:view:9:4: Now you can provide attr `wx:key` for a `wx:for` to improve performance.
[WXML Runtime warning] ./pages/shop/shop.wxml
 Now you can provide attr `wx:key` for a `wx:for` to improve performance.
   7 | </view>
   8 | <scroll-view scroll-y='true' style='height:{{app.windowHeight-30}}px'>
 > 9 |   <view wx:for="{{shangpin}}" class='content' data-shangpincode='{{item.shangpincode}}' bindtap='choose'>
     |   ^
  10 |     <image class='shangpin_img' mode="aspectFit" src="{{url}}{{item.img}}" />
  11 |     <view class='context_text' style='width:{{app.windowWidth-150}}px'>
  12 |       <view style="height:50px">
```

图 8-10

需要增加 wx:key="*this"。修改后的 shop.wxml 程序代码如下。

```
<view class='title_view'>
  <button class='title' type='primary' size='mini' disabled='{{10==typeCode}}' data-arg='10' bindtap="selectTitle">百货</button>
  <button class='title' type='primary' size='mini' disabled='{{20==typeCode}}' data-arg='20' bindtap="selectTitle">餐厨</button>
  <button class='title' type='primary' size='mini' disabled='{{30==typeCode}}' data-arg='30' bindtap="selectTitle">办公</button>
  <button class='title' type='primary' size='mini' disabled='{{40==typeCode}}' data-arg='40' bindtap="selectTitle">游戏</button>
  <button class='title' type='primary' size='mini' disabled='{{50==typeCode}}' data-arg='50' bindtap="selectTitle">动漫</button>
</view>
<scroll-view scroll-y='true' style='height:{{app.windowHeight-30}}px'>
  <view wx:for="{{shangpin}}" class='content' data-shangpincode='{{item.shangpincode}}' bindtap='choose' wx:key="*this">
    <image class='shangpin_img' mode="aspectFit" src="{{url}}{{item.img}}" />
    <view class='context_text' style='width:{{app.windowWidth-150}}px'>
      <view style="height:50px">
        <text class='logo' wx:if='{{""!=item.logo}}'>{{item.logo}}</text>
        <text class="text">{{item.shangpinname}}</text>
      </view>
      <view style="height:20px;font-size:small">
        <text class='biaoqian' wx:if='{{""!=item.baoyou}}'>{{item.baoyou}}</text>
        <text class='biaoqian' wx:if='{{""!=item.gouwuquan}}'>{{item.gouwuquan}}</text>
      </view>
      <view class="price">
        ¥: {{item.price}}
      </view>
    </view>
  </view>
</scroll-view>
```

shop.wxss 的相关程序代码如下。

```
.title_view{
  border-bottom-style:solid;
  border-bottom-color:blue;
  width: 100%;height:30px;
  background-color: #eee;
```

```
}

.title{
  margin-left: 1px;
  margin-right: 1px;
  height:29px;
}

.content{
  display: flex;
  margin: 3px 1px 3px 1px;
  border-bottom: 1px dashed black;
}

.shangpin_img{
  width: 100px;
  height: 100px;
}

.context_text{
  padding: 5px
}

.logo{
  background-color:red;
  color:yellow;
  font-size:small;
  margin-top: 1px;
  width: 2ex;
}

.text{
  font-size:small
}

.biaoqian{
  color:red;
  border-color:red;
  border-style:solid;
  border-width: 1px;
  border-radius: 5px;
  margin: 1px;
  padding: 1px;
  font-size:small;
}

.price{
  color:red;
  font-size:small;
  height:20px;
  bottom:0;
}
```

8.4　“商品详情展示”页面

“商品详情展示”页面中，当用户单击商品列表中的某个商品时，会显示该商品的详细信息。要实现该功能，需要在 shop.js 中实现 choose 函数。

```
  choose: function (event) {

    for (var x = 0; x < this.data.shangpin.length; x++) {
      if (event.currentTarget.dataset.shangpincode == this.data.shangpin[x].shangpin
code) {
        wx.setStorageSync('chooseYigouShangpin', this.data.shangpin[x]);
      }
    }

    wx.navigateTo({ url: '/pages/shop1/shop1?shangpincode=' + event.currentTarget.da
taset.shangpincode });
  }
```

choose 函数的作用是根据用户单击的商品编号（shangpincode）遍历商品集合，先将与其一致的商品信息记录到 chooseYigouShangpin 中，然后跳转到/pages/shop1/shop1 页面，并传递参数 shangpincode。

创建的 shop1 页面如图 8-11 所示。

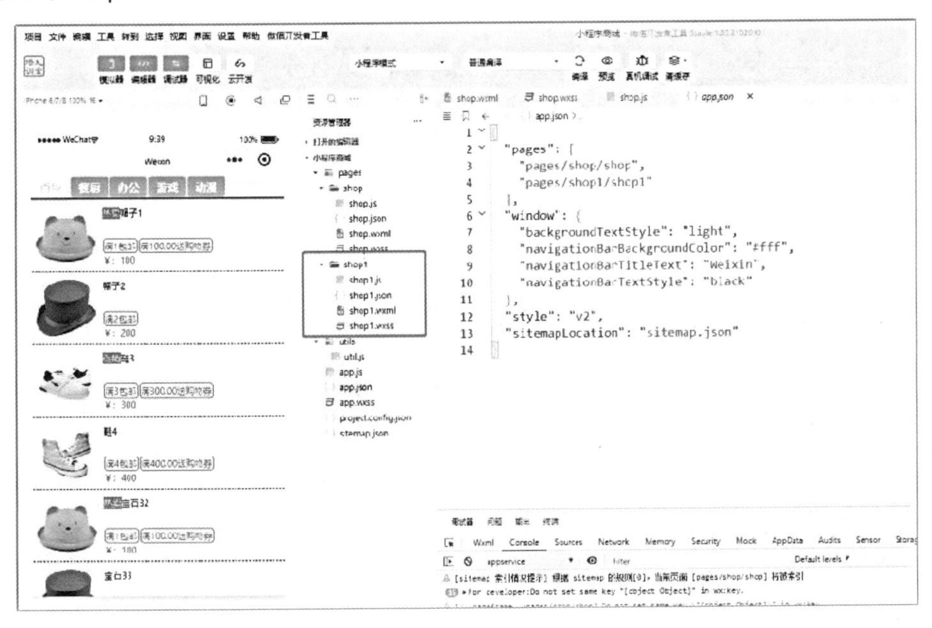

图 8-11

▶ **注意：**

app.json 的配置要相应改变，增加关于 shop1 的配置 pages/shop1/shop1。

为了便于开发，此处不使用普通编译模式，而是使用自定义编译条件。参照图 8-12、图 8-13 进行配置，即可直接访问 shop1 页面。

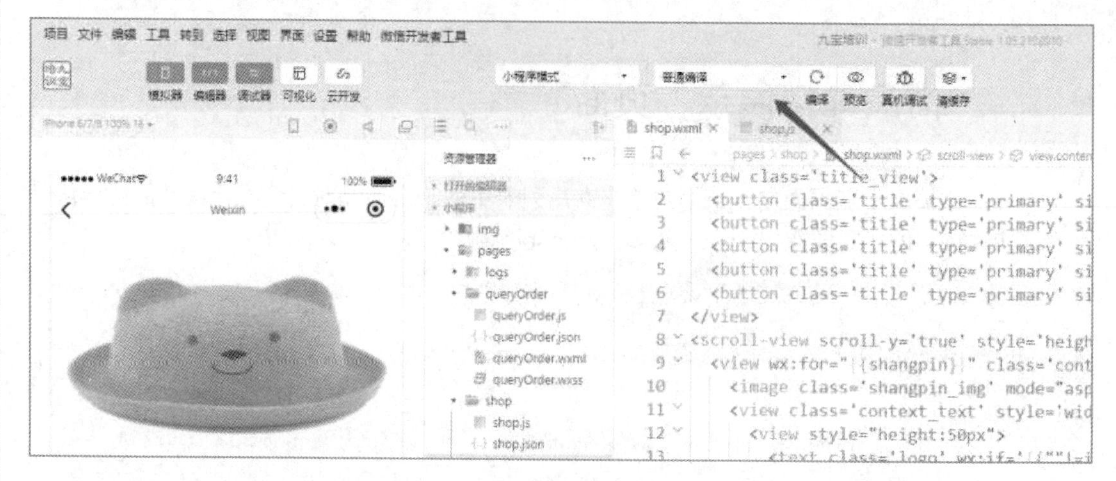

图 8-12

图 8-13

编译条件配置后，即可便捷地开发 shop1。

在 shop1.js 中引用 util.js，代码如下。

```
var util = require('../../utils/util.js');
```

shop1.js 需要实现的程序代码如下。

```
this.setData({ shangpincode: options.shangpincode });
    this.setData({ url: getApp().url });

    util.requestData(this.data.url + 'Shop', { type: 'shopInfo', shangpincode: opt
ions.shangpincode }, function (res) {
      var page = getCurrentPages()[getCurrentPages().length - 1];
      page.setData({ ShangpinXiangqing: res.data });
    });

    var flag = false;

    var YigouShangpin = wx.getStorageSync('YigouShangpin');
    for (var x = 0; x < YigouShangpin.length; x++) {
```

```
        if (this.data.shangpincode == YigouShangpin[x].shangpincode) {
          this.setData({ count: YigouShangpin[x].count });
          flag = true;
          break;
        }
      }

      if (flag) {
        this.setData({ but_text: '修改购买数量' });
      }
```

下面详细讲解 shop1 需要完成的几项工作。

（1）获取 shop 页面传来的参数 shangpincode，记录到变量 shangpincode 中。

（2）从全局变量中得到 url，记录到变量 url 中。

对应程序代码如下。

```
    this.setData({ shangpincode: options.shangpincode });
    this.setData({ url: getApp().url });
```

执行网络请求，向服务器请求的参数是 shangpincode 的商品详情。

```
    util.requestData(this.data.url + 'Shop', { type: 'shopInfo', shangpincode: options
.shangpincode }, function (res) {
      var page = getCurrentPages()[getCurrentPages().length - 1];
      page.setData({ ShangpinXiangqing: res.data });
    });
```

服务器执行的 SQL 代码如下。

```
    select id, typecode, typename, shangpincode, shangpinname, text, nick_name, avatar_url,
createdate from shangpinlog WHERE ( shangpincode = ? )
    DEBUG [http-nio-80-exec-4] - ==> Parameters: 101(String)
```

服务器返回的结果被保存到 ShangpinXiangqing 中，如图 8-14 所示。

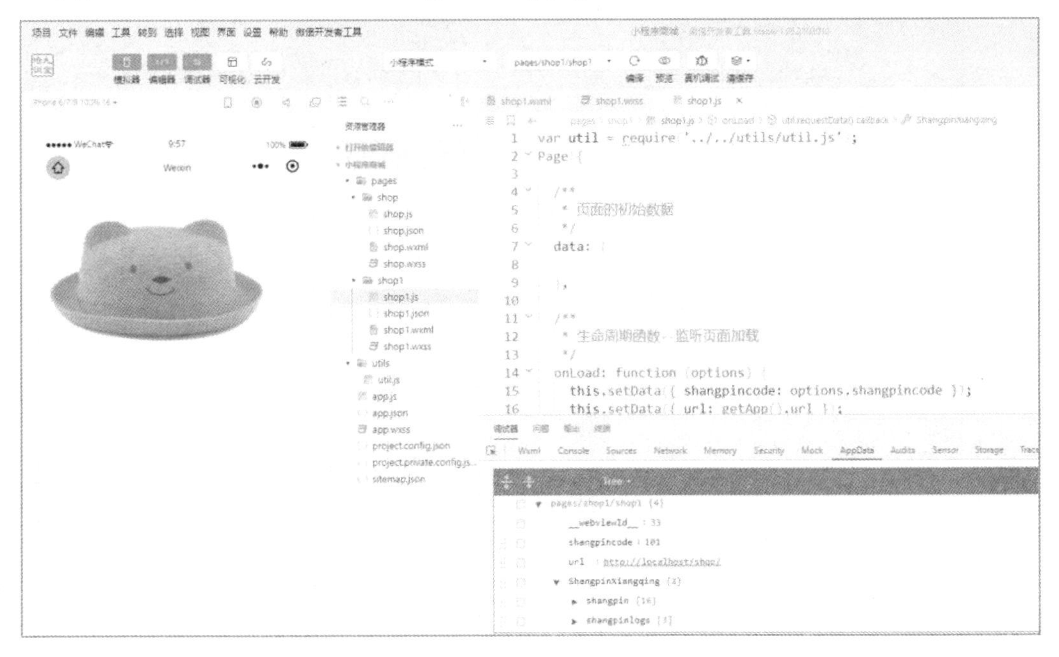

图 8-14

（3）定义购买商品按钮的文字提示。当用户未选择商品时，提示文字为“立即购买”；当用户已选定商品时，提示文字为“修改购买数量”。

（4）当用户选择某商品时，自动回显已选择的购买数量。

程序代码如下，页面效果如图 8-15、图 8-16 所示。

```
var flag = false;

  var YigouShangpin = wx.getStorageSync('YigouShangpin');
  for (var x = 0; x < YigouShangpin.length; x++) {
    if (this.data.shangpincode == YigouShangpin[x].shangpincode) {
      this.setData({ count: YigouShangpin[x].count });
      flag = true;
      break;
    }
  }

  if (flag) {
    this.setData({ but_text: '修改购买数量' });
}
```

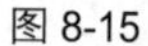
图 8-15

图 8-16

shop1.js 的变量信息如下。

```
data: {
    shangpincode: null,

    ShangpinXiangqing: null,

    url: null,

    /**购买数 */
    count: 1,
```

```
    /**but_text */
    but_text: '立即购买',
  },
```

shop1.wxml 程序代码如下。

```
<view>
  <view wx:if='{{null!=ShangpinXiangqing}}'>
    <image class='shangpin_img' src='{{url}}{{ShangpinXiangqing.shangpin.img}}'></
image>
  </view>
  <view>
    <text class='logo' wx:if='{{""!=ShangpinXiangqing.shangpin.logo}}'>{{ShangpinX
iangqing.shangpin.logo}}</text>
    <text class="shangpinname">{{ShangpinXiangqing.shangpin.shangpinname}}</text>
  </view>
  <view>
    <text class='text'>{{ShangpinXiangqing.shangpin.text }}</text>
  </view>
  <view>
    <text class='biaoqian' wx:if='{{""!=ShangpinXiangqing.shangpin.baoyou}}'>{{Sha
ngpinXiangqing.shangpin.baoyou}}</text>
    <text class='biaoqian' wx:if='{{""!=ShangpinXiangqing.shangpin.gouwuquan}}'>{{
ShangpinXiangqing.shangpin.gouwuquan}}</text>
  </view>
  <view style='font-size:small;'>
    <block wx:if='{{""!=ShangpinXiangqing.shangpin.zengpinbaozheng}}'>
      <icon size='15' type="success" />
      <text>{{ShangpinXiangqing.shangpin.zengpinbaozheng}}</text>
    </block>
    <block wx:if='{{""!=ShangpinXiangqing.shangpin.gouwuquan}}'>
      <icon size='15' type="success" />
      <text>{{ShangpinXiangqing.shangpin.gouwuquan}}</text>
    </block>
    <block wx:if='{{""!=ShangpinXiangqing.shangpin.quanguolianbao}}'>
      <icon size='15' type="success" />
      <text>{{ShangpinXiangqing.shangpin.quanguolianbao }}</text>
    </block>
    <block wx:if='{{""!=ShangpinXiangqing.shangpin.zengyunfeixian}}'>
      <icon size='15' type="success" />
      <text>{{ShangpinXiangqing.shangpin.zengyunfeixian }}</text>
    </block>
  </view>
  <view class="price">
    ￥: {{ShangpinXiangqing.shangpin.price}}
  </view>
  <view style='display: flex;'>
    <button size='mini' bindtap='subtraction'>-</button>
    <text>{{count}}</text>
    <button size='mini' bindtap='addition'>+</button>
    <button size='mini' type='primary' bindtap='buy'>{{but_text}}</button>
  </view>
  <view style='margin-top:10px'>
    <text style='font-weight:bold'>买家评价</text>
    <view wx:for='{{ShangpinXiangqing.shangpinlogs}}' style='background-color:{{(0
==index%2)?"antiquewhite":"burlywood"}};margin-top:10px;' wx:for-item='log' wx:for-ind
```

```
ex='index'>
          <view style='display: flex;'>
            <image mode='aspectFit' style='width:25px;height:15px' src='{{url}}{{log.a
vatarUrl}}'></image>
            <text style='font-size:small;width:30%'>{{log.nickName}}</text>
            <view style='width:100%'></view>
            <text style='font-size:small;width:50%;'>{{log.createdate}}</text>
          </view>
          <view>
            <text>{{log.text}}</text>
          </view>
        </view>
      </view>
    </view>
```

现在分段讲解这部分程序。

（1）实现商品详情图片展示，程序代码如下。

```
    <view wx:if='{{null!=ShangpinXiangqing}}'>
      <image class='shangpin_img' src='{{url}}{{ShangpinXiangqing.shangpin.img}}'></im
age>
    </view>
```

注意：

之所以增加 wx:if='{{null!=ShangpinXiangqing}}'是因为渲染与加载数据的顺序问题。由于执行的是异步网络请求，当服务端请求的数据没有响应到微信小程序时，微信小程序的本地渲染可能已经结束。因此，微信小程序开发工具的调试器出现提示，如图 8-17 所示。

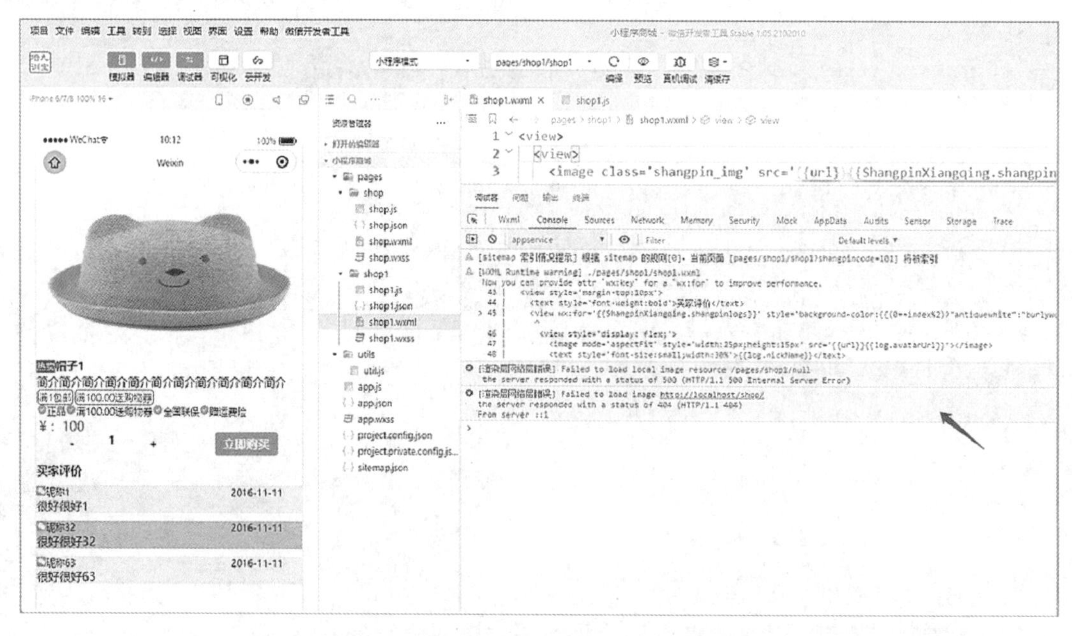

图 8-17

（2）实现显示商品的文字信息，相关程序如下，效果如图 8-18 所示。

```
    <view>
      <text class='text'>{{ShangpinXiangqing.shangpin.text }}</text>
```

```
    </view>
    <view>
      <text class='biaoqian' wx:if='{{""!=ShangpinXiangqing.shangpin.baoyou}}'>{{Shang
pinXiangqing.shangpin.baoyou}}</text>
      <text class='biaoqian' wx:if='{{""!=ShangpinXiangqing.shangpin.gouwuquan}}'>{{Sh
angpinXiangqing.shangpin.gouwuquan}}</text>
    </view>
    <view style='font-size:small;'>
      <block wx:if='{{""!=ShangpinXiangqing.shangpin.zengpinbaozheng}}'>
        <icon size='15' type="success" />
        <text>{{ShangpinXiangqing.shangpin.zengpinbaozheng}}</text>
      </block>
      <block wx:if='{{""!=ShangpinXiangqing.shangpin.gouwuquan}}'>
        <icon size='15' type="success" />
        <text>{{ShangpinXiangqing.shangpin.gouwuquan}}</text>
      </block>
      <block wx:if='{{""!=ShangpinXiangqing.shangpin.quanguolianbao}}'>
        <icon size='15' type="success" />
        <text>{{ShangpinXiangqing.shangpin.quanguolianbao }}</text>
      </block>
      <block wx:if='{{""!=ShangpinXiangqing.shangpin.zengyunfeixian}}'>
        <icon size='15' type="success" />
        <text>{{ShangpinXiangqing.shangpin.zengyunfeixian }}</text>
      </block>
    </view>
```

（3）实现显示商品价格，相关程序如下，效果如图 8-19 所示。

```
    <view class="price">
        ¥: {{ShangpinXiangqing.shangpin.price}}
      </view>
```

图 8-18

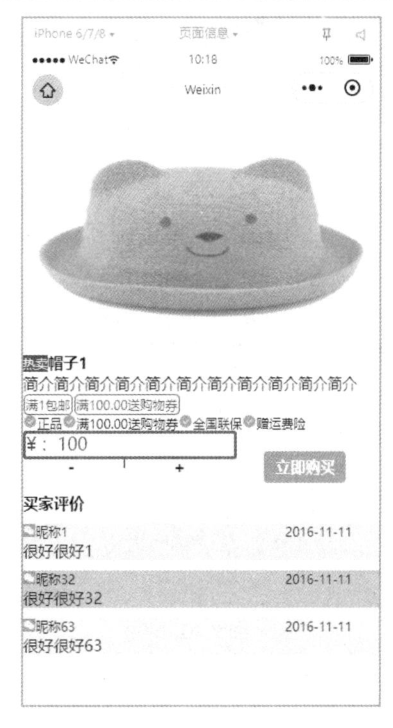

图 8-19

（4）实现确定购买商品数量，相关程序如下，效果如图 8-20 所示。

```
<view style='display: flex;'>
    <button size='mini' bindtap='subtraction'>-</button>
    <text>{{count}}</text>
    <button size='mini' bindtap='addition'>+</button>
    <button size='mini' type='primary' bindtap='buy'>{{but_text}}</button>
</view>
```

（5）实现买家评价，相关程序如下，效果如图 8-21 所示。

```
<view style='margin-top:10px'>
    <text style='font-weight:bold'>买家评价</text>
    <view wx:for='{{ShangpinXiangqing.shangpinlogs}}' style='background-color:{{(0==index%2)?"antiquewhite":"burlywood"}};margin-top:10px;' wx:for-item='log' wx:for-index='index'>
      <view style='display: flex;'>
        <image mode='aspectFit' style='width:25px;height:15px' src='{{url}}{{log.avatarUrl}}'></image>
        <text style='font-size:small;width:30%'>{{log.nickName}}</text>
        <view style='width:100%'></view>
        <text style='font-size:small;width:50%;'>{{log.createdate}}</text>
      </view>
      <view>
        <text>{{log.text}}</text>
      </view>
    </view>
  </view>
```

图 8-20

图 8-21

对于 shop1，开发时需要注意购买商品数量相关的逻辑。主要有以下几点。

◆ 对于不同的商品，购买的数量不同。

◆　已下单到购物车的商品回显时，需要同步回显购物车中该商品的下单数量。

◆　单击“+”或“-”按钮，可以增加或减少相应商品数量。

对应地，需要定义 subtraction 函数与 addition 函数，相关程序如下。

```
addition: function (event) {

  this.setData({ count: this.data.count * 1 + 1 });

},
subtraction: function (event) {

  if (0 > this.data.count * 1 - 1) {
    this.setData({ count: 0 });
  } else {
    this.setData({ count: this.data.count * 1 - 1 });
  }

},
```

注意：

设置“-”按钮时，当购买数小于 0 时，不能出现负数。

变量 count 的作用是显示该商品的下单数量。修改当前商品的下单数量，相应变量中记录的该商品的数量要同步修改，如图 8-22 所示。

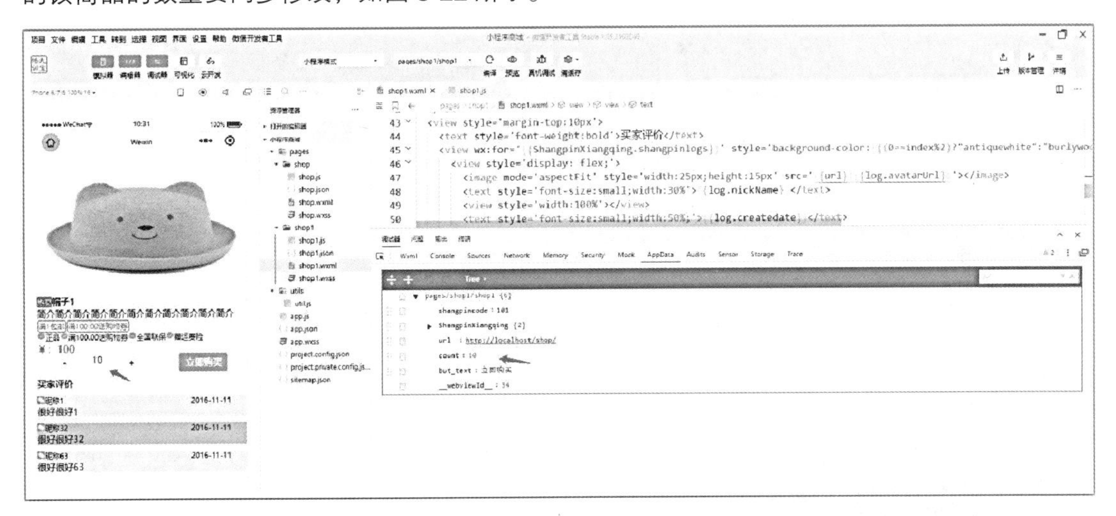

图 8-22

此时，修改目标商品。单击“编译模式”按钮，如图 8-23 所示，在下拉菜单中选择“添加编译模式”命令，如图 8-24 所示，在打开的“添加编译模式”对话框中进行设置，如图 8-25 所示。

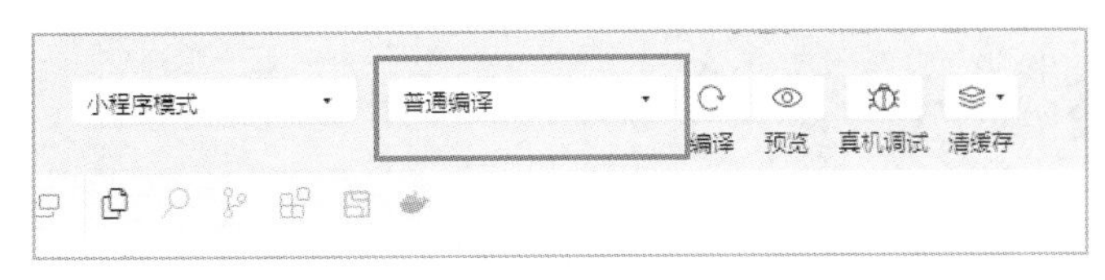

图 8-23

小程序模式
普通编译
真机调试
清缓存
普通编译
添加编译模式
快速 二维码编译
快速 URL Scheme 编译
快速 URL Link 编译
快速 Short Link 编译
资源管理器
打开的编辑器
小程序
img
pages
utils
app.js
app.json
app.wxss

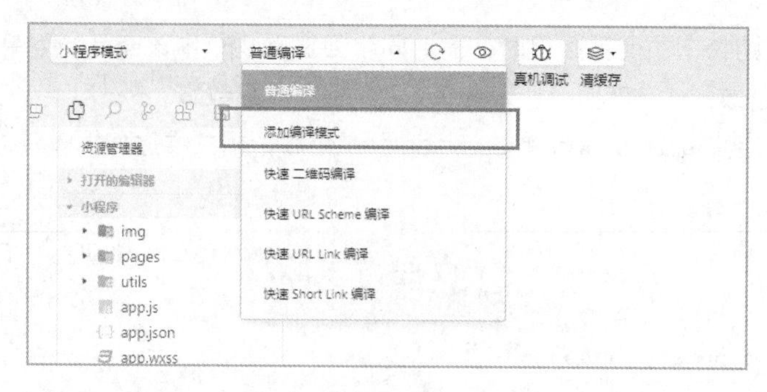

图 8-24

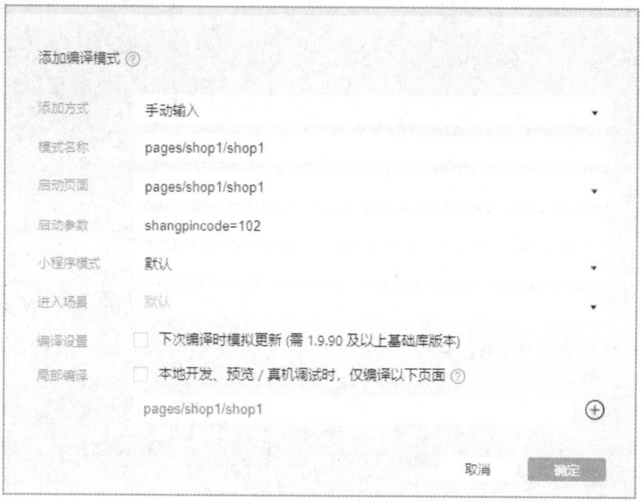

添加编译模式
添加方式
手动输入
模式名称
pages/shop1/shop1
启动页面
pages/shop1/shop1
启动参数
shangpincode=102
小程序模式
默认
进入场景
默认
编译设置
下次编译时模拟更新 (需 1.9.90 及以上基础库版本)
局部编译
本地开发、预览 / 真机调试时，仅编译以下页面
pages/shop1/shop1
取消
确定

图 8-25

再次编译微信小程序，将显示 shangpincode =102 的商品，如图 8-26 所示。

帽子2
买家评价
立即购买
shangpincode : 102
url : http://localhost/shop/
but_text : 立即购买

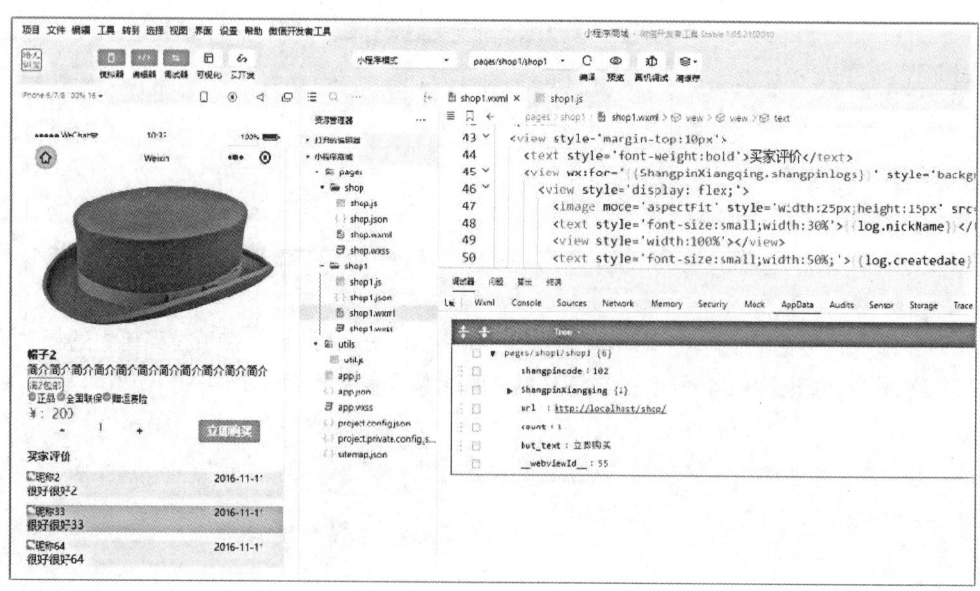

图 8-26

同样，单击模拟器中商品数量位置的“+”或“-”按钮，查看本地记录的商品信息变化，如图 8-27 所示。

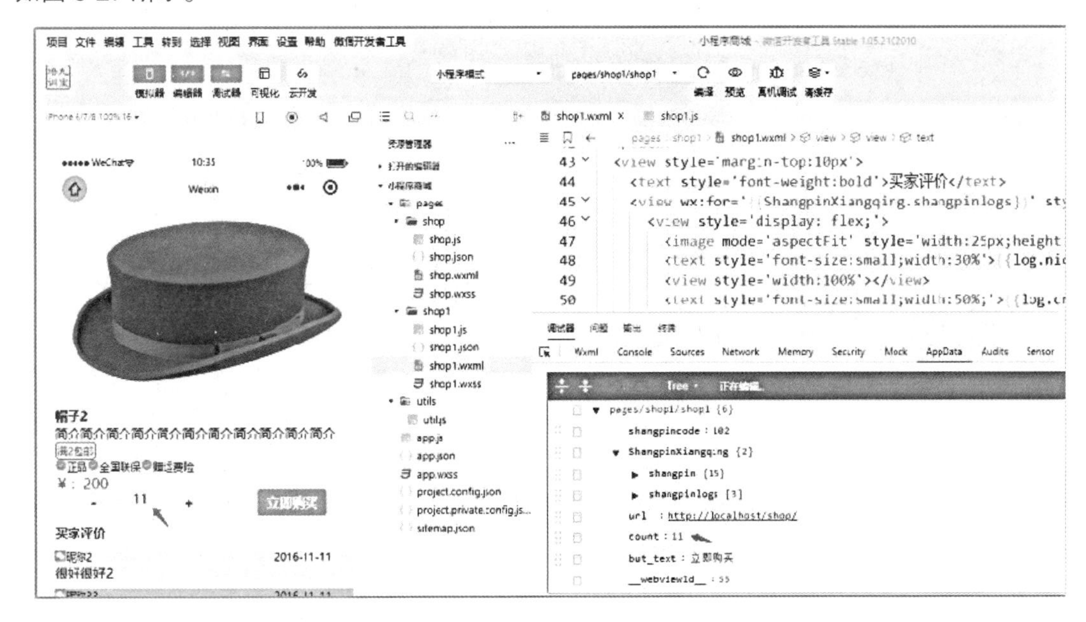

图 8-27

用户单击“立即购买”和“修改购买数量”按钮时，会将当前商品下单信息记录到购物车中，相关程序如下。

```
buy: function () {

  var YigouShangpin = wx.getStorageSync('YigouShangpin');

  var flag = false;
  for (var x = 0; x < YigouShangpin.length; x++) {
    if (this.data.shangpincode == YigouShangpin[x].shangpincode) {
      YigouShangpin[x].count = this.data.count;
      flag = true;
      break;
    }
  }

  if (!flag) {
    var arg = {};
    arg = wx.getStorageSync('chooseYigouShangpin');
    arg.count = this.data.count;
    YigouShangpin.push(arg);
    this.setData({ count: arg.count });
  }

  if (0 == this.data.count) {
    this.setData({ but_text: '立即购买' });
  } else {
    this.setData({ but_text: '修改购买数量' });
```

```
    }
    YigouShangpin: wx.setStorageSync('YigouShangpin', YigouShangpin);
}
```

▶ **注意：**

应该先清除本地缓存，如图 8-28 所示。

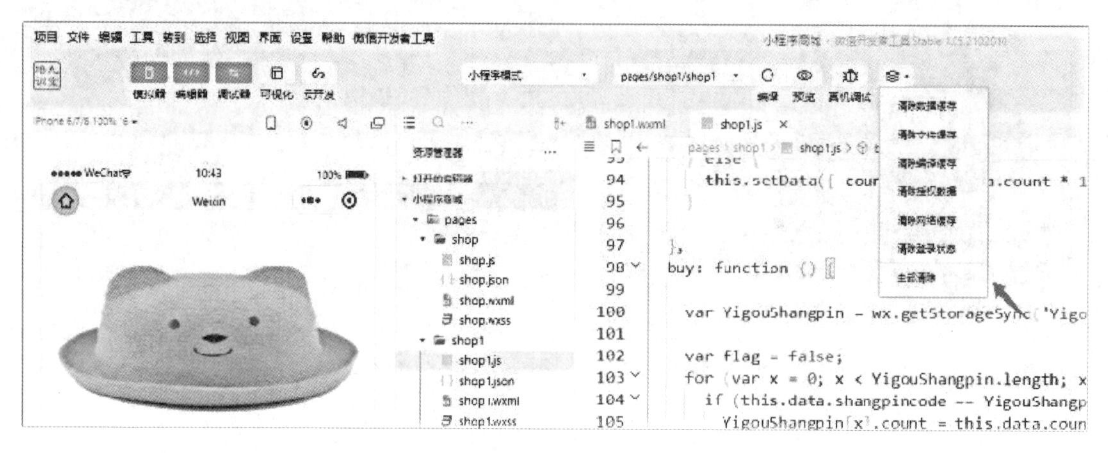

图 8-28

编译 shop 页面，可以看到本地缓存的变量 YigouShangpin，如图 8-29 所示。

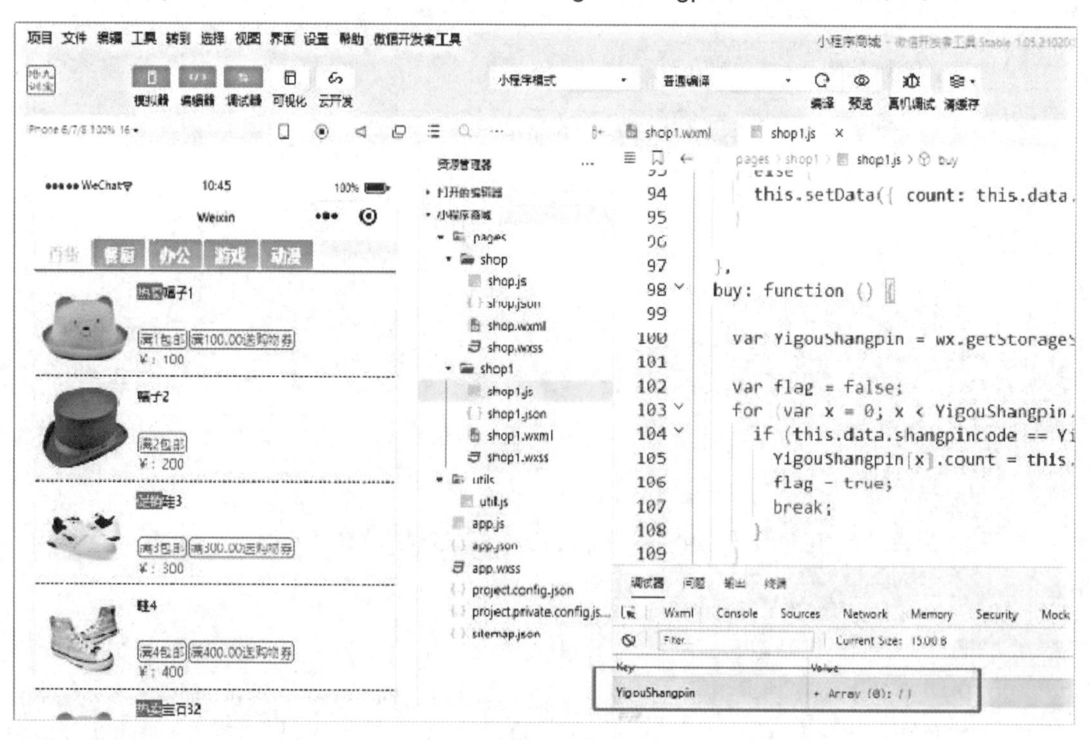

图 8-29

选择 shangpincode =101 的商品，如图 8-30 所示。

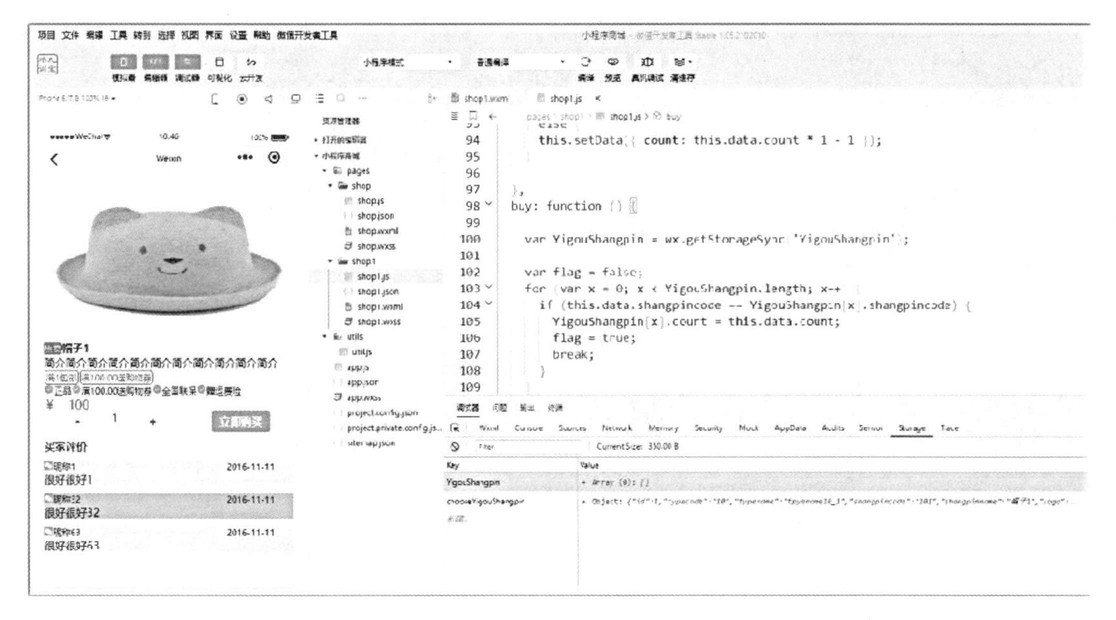

图 8-30

先修改购买数量为 5，然后单击“立即购买”按钮，如图 8-31 所示。

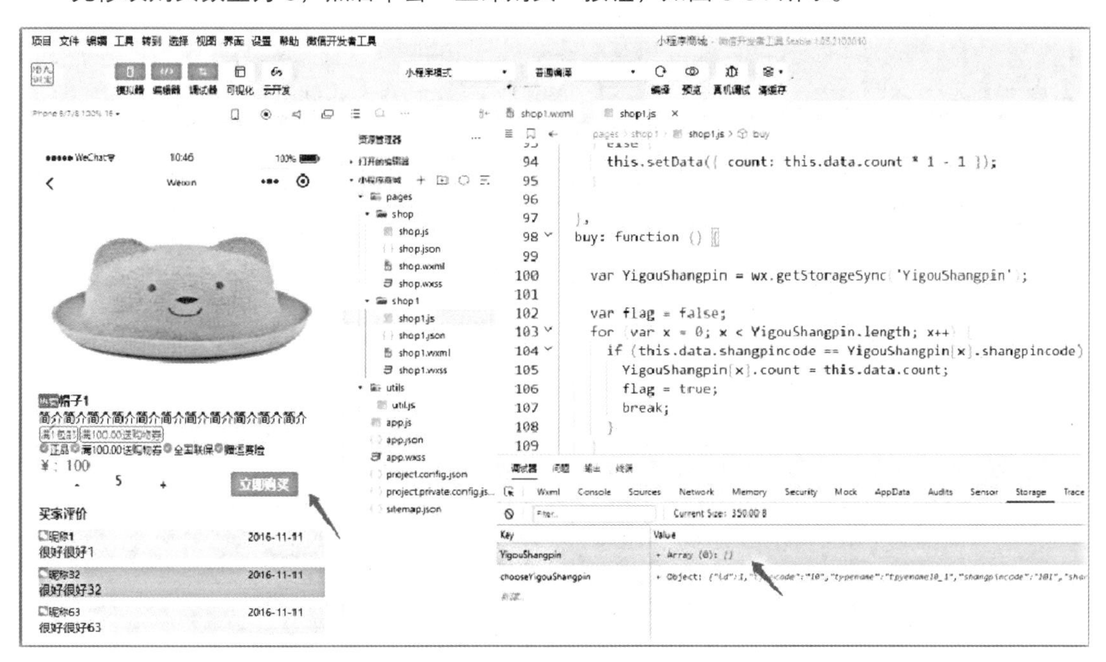

图 8-31

回退，选择 shangpincode =102 的商品，如图 8-32 所示。

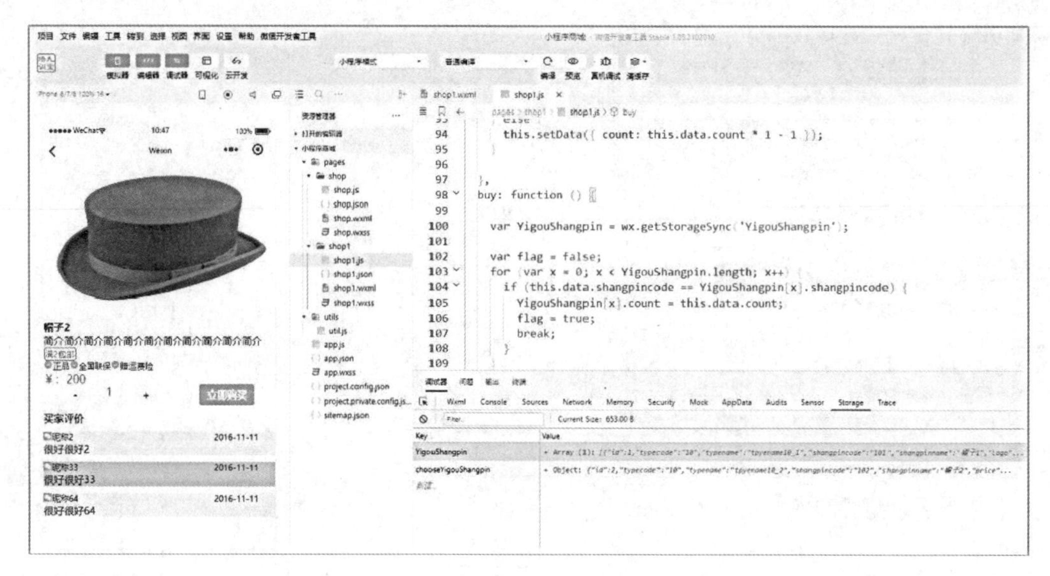

图 8-32

效仿上述测试，先修改购买数量，然后单击“立即购买”按钮，如图 8-33 所示。

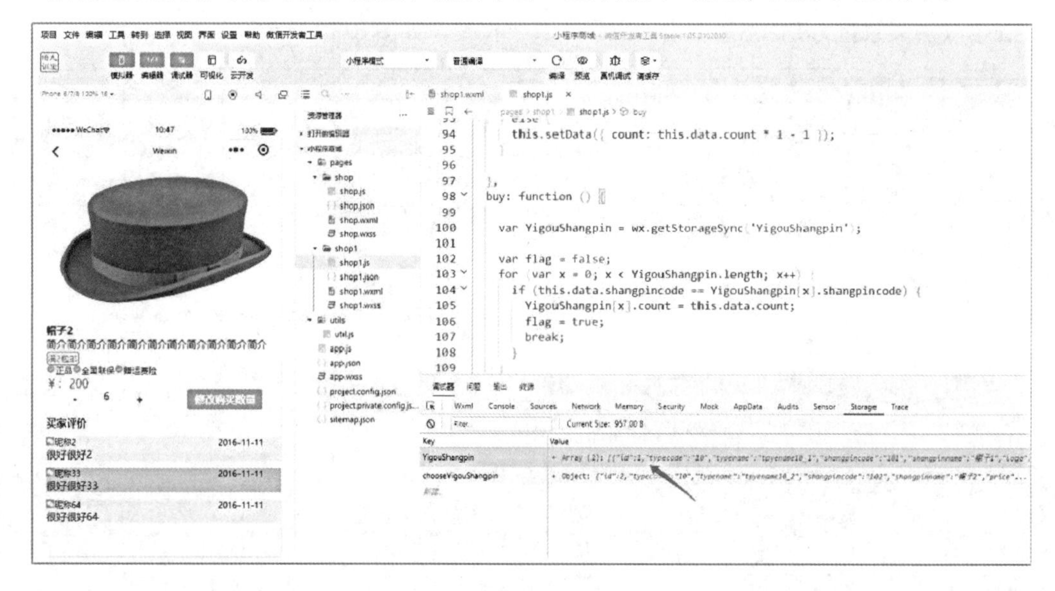

图 8-33

此时，购物车变量 YigouShangpin 发生了改变。

shop1.wxss 的相关程序代码如下。

```
.shangpin_img{
  width: 100%;
  margin: 0;
  padding: 0;
}

.logo{
  background-color:red;
  color:yellow;
  font-size:small;
  margin-top: 1px;
```

```
    width: 2ex;
  }

  .shangpinname{
    font-weight:bold
  }

  .text{

  }

  .biaoqian{
    color:red;
    border-color:red;
    border-style:solid;
    border-width: 1px;
    border-radius: 5px;
    margin: 1px;
   padding: 1px;
    font-size:small;
  }

  .price{
    color:red;
    font-size:large;
    height:20px;
    bottom:0;
  }
```

注意：

本节演示的是简化后的教学案例，在实际的小程序商城系统开发中，还应该增加“商品库存数”等信息，以及“下单商品数”不超过“商品库存数”等逻辑约束。

8.5　tabBar

微信小程序通常是多 tab 应用，因此微信客户端窗口的底部或顶部会有一个 tab 栏，以供用户切换页面。通过 tabBar 配置项，可以指定 tab 栏和 tab 切换时显示的页面。

tabBar 属性如表 8-1 所示。

表 8-1

属　性	类　型	必　填	默 认 值	描　述	最 低 版 本
color	HexColor	是		tab 上的文字默认颜色，仅支持十六进制颜色	
selectedColor	HexColor	是		tab 上的文字选中时的颜色，仅支持十六进制颜色	
backgroundColor	HexColor	是		tab 的背景色，仅支持十六进制颜色	
borderStyle	string	否	black	tabBar 上边框的颜色，仅支持 black / white	
list	Array	是		tab 的列表，详见表 8-2，最少 2 个、最多 5 个 tab	
position	string	否	bottom	tabBar 的位置，仅支持 bottom / top	
custom	boolean	否	false	自定义 tabBar	2.5.0

list 是一个数组，其中可配置 2 ~ 5 个 tab。tab 按数组的顺序排列，每项都是一个对象，其属性值如表 8-2 所示。

表 8-2

属　性	类　型	必　填	说　明
pagePath	string	是	页面路径，必须在 pages 中先定义
text	string	是	tab 上按钮文字
iconPath	string	否	图片路径，icon 大小限制为 40 KB，建议尺寸为 81px × 81px，不支持网络图片。当 position 为 top 时，不显示 icon
selectedIconPath	string	否	选中时的图片路径，icon 大小限制为 40 KB，建议尺寸为 81px × 81px，不支持网络图片。当 position 为 top 时，不显示 icon

修改 app.json，定义 tabBar 的代码如下。

```
"tabBar": {
  "list": [{
    "pagePath": "pages/shop/shop",
    "text": "卖场",
    "iconPath":"img/gwc.jpg",
    "selectedIconPath":"img/gwc.jpg"
  }, {
    "pagePath": "pages/queryOrder/queryOrder",
    "text": "查看购物单",
    "iconPath":"img/pay.jpg",
    "selectedIconPath":"img/pay.jpg"
  }]
},
```

注意：

需要先创建 queryOrder 文件夹，并在该文件夹下创建 queryOrder 页面。

本案例要实现的 tabBar 如图 8-34 所示。

图 8-34

8.6 “购物车”页面

queryOrder 实现下单商品展示与物流信息采集等功能。queryOrder.js 需要先引用“var util = require('../../utils/util.js');”。

onLoad 事件用于获取全局变量 url，相关程序如下。

```
this.setData({ url: getApp().url });
```

onShow 事件用于读取本地变量 YigouShangpin 中的新订单信息，并计算订单总价，相关代码如下。

```
this.setData({ YigouShangpin: wx.getStorageSync('YigouShangpin') });
  var all_price = 0;
  for (var x = 0; x < this.data.YigouShangpin.length; x++) {
    all_price += this.data.YigouShangpin[x].price * this.data.YigouShangpin[x].cou
nt * 100;
  }
  this.setData({ all_price: all_price / 100 });
```

需要定义的变量如下。

```
data: {
  YigouShangpin: null,
  all_price: 0,
  userInfo: null,
  url: ''
},
```

queryOrder.wxml 的程序代码如下。

```
<view>
  <view style="display: flex;">
    <view class="shangpinname">商品名</view>
    <view class="price">单价</view>
    <view class="count">数量</view>
  </view>

  <view style="display: flex;" wx:if="{{10==item.typecode}}" class="typecode{{item
.typecode}}" wx:for="{{YigouShangpin}}" wx:key="*this">
    <view class="shangpinname">{{item.shangpinname}}</view>
    <view class="price">{{item.price}}</view>
    <view class="count">{{item.count}}</view>
  </view>

  <view style="display: flex;" wx:if="{{20==item.typecode}}" class="typecode{{item
.typecode}}" wx:for="{{YigouShangpin}}" wx:key="*this">
    <view class="shangpinname">{{item.shangpinname}}</view>
    <view class="price">{{item.price}}</view>
    <view class="count">{{item.count}}</view>
  </view>

  <view style="display: flex;" wx:if="{{30==item.typecode}}" class="typecode{{item
```

```
.typecode}}" wx:for="{{YigouShangpin}}" wx:key="*this">
        <view class="shangpinname">{{item.shangpinname}}</view>
        <view class="price">{{item.price}}</view>
        <view class="count">{{item.count}}</view>
        </view>

      <view style="display: flex;" wx:if="{{40==item.typecode}}" class="typecode{{item
.typecode}}" wx:for="{{YigouShangpin}}" wx:key="*this">
        <view class="shangpinname">{{item.shangpinname}}</view>
        <view class="price">{{item.price}}</view>
        <view class="count">{{item.count}}</view>
      </view>

      <view style="display: flex;" wx:if="{{50==item.typecode}}" class="typecode{{item
.typecode}}" wx:for="{{YigouShangpin}}" wx:key="*this">
        <view class="shangpinname">{{item.shangpinname}}</view>
        <view class="price">{{item.price}}</view>
        <view class="count">{{item.count}}</view>
      </view>

      <view>合计: ¥{{all_price}}</view>

      <view class="form">

        <form bindsubmit="formSubmit">

        <view class="element">
          <label>收件人:
          <input type="text" name="sjr" placeholder="收件人" value="收件人"/></label>
        </view>

        <view class="element">
          <label>邮递地址:
          <textarea name="dz" placeholder="邮递地址" value="邮递地址"/></label>
        </view>

        <view class="element">
          <label>联系电话:
          <input name="dh" type="idcard" placeholder="联系电话" value="联系电话
"/></label>
        </view>

        <view style="display: flex; border-top-style:solid; border-top-color:blue;padd
ing-top:10px;">
          <button type="primary" form-type="submit" size="mini" bindtap="submit"> subm
it </button>
          <button type="warn"    form-type="reset"  size="mini" bindtap="reset" > rese
t </button>
        </view>

        </form>

      </view>

    </view>
```

queryOrder.wxml 实现的几个功能如下。

（1）展示购物车中的商品，需要提供商品名、单价、数量。

要求商品按照类别展示，因此需要通过 wx:if="{{10==item.typecode}}"进行判断。相关程序代码如下，实现页面如图 8-35 所示。

```
<view style="display: flex;">
  <view class="shangpinname">商品名</view>
  <view class="price">单价</view>
  <view class="count">数量</view>
</view>

<view style="display: flex;" wx:if="{{10==item.typecode}}" class="typecode{{item.t
ypecode}}" wx:for="{{YigouShangpin}}" wx:key="*this">
  <view class="shangpinname">{{item.shangpinname}}</view>
  <view class="price">{{item.price}}</view>
  <view class="count">{{item.count}}</view>
</view>

<view style="display: flex;" wx:if="{{20==item.typecode}}" class="typecode{{item.t
ypecode}}" wx:for="{{YigouShangpin}}" wx:key="*this">
  <view class="shangpinname">{{item.shangpinname}}</view>
  <view class="price">{{item.price}}</view>
  <view class="count">{{item.count}}</view>
</view>

<view style="display: flex;" wx:if="{{30==item.typecode}}" class="typecode{{item.t
ypecode}}" wx:for="{{YigouShangpin}}" wx:key="*this">
  <view class="shangpinname">{{item.shangpinname}}</view>
  <view class="price">{{item.price}}</view>
  <view class="count">{{item.count}}</view>
</view>

<view style="display: flex;" wx:if="{{40==item.typecode}}" class="typecode{{item.t
ypecode}}" wx:for="{{YigouShangpin}}" wx:key="*this">
  <view class="shangpinname">{{item.shangpinname}}</view>
  <view class="price">{{item.price}}</view>
  <view class="count">{{item.count}}</view>
</view>
```

（2）自动计算购物车商品总价，程序代码如下，实现页面如图 8-36 所示。

```
<view>合计: ￥{{all_price}}</view>
```

注意：

在实际小程序商城开发中，订单总价建议以服务端计算结果为准。

（3）获取物流信息，包括收件人、邮递地址和联系电话，程序代码如下。

```
<view class="form">

  <form bindsubmit="formSubmit">

  <view class="element">
    <label>收件人:
```

```
      <input type="text" name="sjr" placeholder="收件人" value="收件人"/></label>
    </view>

    <view class="element">
      <label>邮递地址:
      <textarea name="dz" placeholder="邮递地址" value="邮递地址"/></label>
    </view>

    <view class="element">
      <label>联系电话:
      <input name="dh" type="idcard" placeholder="联系电话" value="联系电话"/></label>
    </view>

    <view style="display: flex; border-top-style:solid; border-top-color:blue;paddin
g-top:10px;">
      <button type="primary" form-type="submit" size="mini" bindtap="submit"> submit
 </button>
      <button type="warn"   form-type="reset"  size="mini" bindtap="reset" > reset
</button>
    </view>

    </form>

  </view>
```

获取物流信息页面如图 8-37 所示。

图 8-35　　图 8-36　　图 8-37

（4）提交 button，代码如下。

```
  <button type="primary" form-type="submit" size="mini" bindtap="submit"> submit </b
utton>
```

关联的函数应该修改 queryOrder.js，代码如下。

```
    onShow: function () {
      this.setData({ YigouShangpin: wx.getStorageSync('YigouShangpin') });
      var all_price = 0;
      for (var x = 0; x < this.data.YigouShangpin.length; x++) {
        all_price += this.data.YigouShangpin[x].price * this.data.YigouShangpin[x].cou
nt * 100;
      }
      this.setData({ all_price: all_price / 100 });
    },
```

（5）重置 button，代码如下。

```
    <button type="warn"    form-type="reset"  size="mini" bindtap="reset" > reset </bu
tton>
```

queryOrder.wxss 代码如下。

```
    .shangpinname {
      font-weight: bold;
      font-size: large;
        width: 100%;
    }

    .price {
      color: red;
      font-weight: bold;
      padding-left: 1rem;
      padding-right: 1rem;
        width: 150px;
    }

    .count {
      padding-left: 10px;
      padding-right: 10px;
      color: blue;
      font-weight: bold;
        width: 150px;
    }

    .typecode10{background-color:antiquewhite;}
    .typecode20{background-color:beige;}
    .typecode30{background-color:gainsboro;}
    .typecode40{background-color:wheat;}
    .typecode50{background-color:darkseagreen;}

    view{
      margin: 10px;
    }

    .element{
      margin-top: 10px;
    }

    .form{
```

```
    border-style:double;
    border-color:chartreuse;
    border-radius:10px;
  }

  input , textarea{
    border-style:solid;
    border-width: 1px;
  }
```

相应地，服务端接收微信小程序提交的数据后，需要记录到数据库中。相关程序代码如下。

```
    String orderId = WX_Util.getSequ("OrderId");

    Date date = new Date();
    SimpleDateFormat simpleDateFormat = new SimpleDateFormat("yyyyMMddHHmmss_");
    orderId = simpleDateFormat.format(date)+orderId;

    boolean flag = ShopService.saveOrder(
                orderId,
                date,
                gson.fromJson(request.getParameter("userInfo"), Userinfo.class),
                gson.fromJson(request.getParameter("userInfo_"), Userinfo.class),
                gson.fromJson(request.getParameter("YigouShangpin"), new TypeToken
<List<Xiaoshou>>(){}.getType()));
    if(flag){
      out.print("{\"flag\":\""+flag+"\",\"orderId\":\""+orderId+"\"}");
    }else{
      out.print("{\"flag\":\""+flag+"\"}");
    }
```

提交的数据需要记录当前用户信息，这可以通过调用 wx.getUserInfo 获取，相关程序代码如下。

```
    onReady: function () {
      wx.getUserInfo({
        success: function (res) {
          var pages = getCurrentPages();
          var page = pages[pages.length - 1];
          page.setData({ userInfo: res.userInfo });
        }
      })
    },
```

注意：

本案例是教学案例，很多功能都做了简化。在实际小程序商城开发中需求更复杂，需要验证订单、实现小程序支付等功能。一般来说，微信小程序支付回调成功时，需要修改订单状态为“已支付”，同时推送“微信小程序支付成功”消息。

本案例中，将订单数据记录到数据库中即可，相关程序代码如下。

```
    String orderId = WX_Util.getSequ("OrderId");

    Date date = new Date();
    SimpleDateFormat simpleDateFormat = new SimpleDateFormat("yyyyMMddHHmmss_");
    orderId = simpleDateFormat.format(date)+orderId;
```

```
    boolean flag = ShopService.saveOrder(
                    orderId,
                    date,
                    gson.fromJson(request.getParameter("userInfo"), Userinfo.class),
                    gson.fromJson(request.getParameter("userInfo_"), Userinfo.class),
                    gson.fromJson(request.getParameter("YigouShangpin"), new TypeToken
<List<Xiaoshou>>(){}.getType()));
    if(flag){
      out.print("{\"flag\":\""+flag+"\",\"orderId\":\""+orderId+"\"}");
    }else{
      out.print("{\"flag\":\""+flag+"\"}");
    }
```

单击 submit 按钮，服务器打印的报文如下。

```
    DEBUG [http-nio-80-exec-7] - Opening JDBC Connection
    DEBUG [http-nio-80-exec-7] - Checked out connection 1865406197 from pool.
    DEBUG [http-nio-80-exec-7] - Setting autocommit to false on JDBC Connection
[com.mysql.cj.jdbc.ConnectionImpl@6f2fd6f5]
    DEBUG [http-nio-80-exec-7] - ==>  Preparing: select id, col_name, code, text from sequ
WHERE ( col_name = ? )
    DEBUG [http-nio-80-exec-7] - ==> Parameters: OrderId(String)
    DEBUG [http-nio-80-exec-7] - <==      Total: 1
    DEBUG [http-nio-80-exec-7] - ==>  Preparing: update sequ set col_name = ?, code = ?,
text = ? where id = ?
    DEBUG [http-nio-80-exec-7] - ==> Parameters: OrderId(String), 41(String), null,
1(Integer)
    DEBUG [http-nio-80-exec-7] - <==    Updates: 1
    DEBUG [http-nio-80-exec-7] - Committing JDBC Connection
[com.mysql.cj.jdbc.ConnectionImpl@6f2fd6f5]
    DEBUG [http-nio-80-exec-7] - Resetting autocommit to true on JDBC Connection
[com.mysql.cj.jdbc.ConnectionImpl@6f2fd6f5]
    DEBUG [http-nio-80-exec-7] - Closing JDBC Connection
[com.mysql.cj.jdbc.ConnectionImpl@6f2fd6f5]
    DEBUG [http-nio-80-exec-7] - Returned connection 1865406197 to pool.
    DEBUG [http-nio-80-exec-7] - Opening JDBC Connection
    DEBUG [http-nio-80-exec-7] - Checked out connection 1865406197 from pool.
    DEBUG [http-nio-80-exec-7] - Setting autocommit to false on JDBC Connection
[com.mysql.cj.jdbc.ConnectionImpl@6f2fd6f5]
    DEBUG [http-nio-80-exec-7] - ==>  Preparing: insert into userinfo ( orderid, nick_name,
avatar_url, gender, province, city, sjr, dz, dh, createtime ) values
( ?, ?, ?, ?, ?, ?, ?, ?, ?, ? )
    DEBUG [http-nio-80-exec-7] - ==> Parameters: 20210410151724_41(String), 微信用户
(String),
https://thirdwx.qlogo.cn/mmopen/vi_32/POgEwh4mIHO4nibHOKlMECNjjGxQUq24ZEaGT4poC6icRicc
VGKSyXwibcPq4BWmiaIGuGlicwxaQX6grC9VemZoJ8rg/132(String), 0(String), (String), (String),
收件人(String), 邮递地址(String), 联系电话(String), 2021-04-10 15:17:24.08(Timestamp)
    DEBUG [http-nio-80-exec-7] - <==    Updates: 1
    DEBUG [http-nio-80-exec-7] - ==>  Preparing: insert into xiaoshou ( orderid, count,
typecode, typename, shangpincode, shangpinname, price, img, text, createtime ) values
( ?, ?, ?, ?, ?, ?, ?, ?, ?, ? )
    DEBUG [http-nio-80-exec-7] - ==> Parameters: 20210410151724_41(String), 1(Integer),
10(String), tpyename10_1(String), 101(String), 帽子1(String), 100(BigDecimal),
img/1001.jpg(String), 简介简介简介简介简介简介简介简介简介简介简介(String), 2021-04-10
15:17:24.08(Timestamp)
    DEBUG [http-nio-80-exec-7] - <==    Updates: 1
```

```
    DEBUG [http-nio-80-exec-7] - ==>  Preparing: insert into xiaoshou ( orderid, count, typecode, typename, shangpincode, shangpinname, price, img, text, createtime ) values ( ?, ?, ?, ?, ?, ?, ?, ?, ?, ? )
    DEBUG [http-nio-80-exec-7] - ==> Parameters: 20210410151724_41(String), 1(Integer), 10(String), tpyename10_2(String), 102(String), 帽子2(String), 200(BigDecimal), img/1002.jpg(String), 简介简介简介简介简介简介简介简介简介简介简介简介(String), 2021-04-10 15:17:24.08(Timestamp)
    DEBUG [http-nio-80-exec-7] - <==    Updates: 1
    DEBUG [http-nio-80-exec-7] - Committing JDBC Connection [com.mysql.cj.jdbc.ConnectionImpl@6f2fd6f5]
    DEBUG [http-nio-80-exec-7] - Resetting autocommit to true on JDBC Connection [com.mysql.cj.jdbc.ConnectionImpl@6f2fd6f5]
    DEBUG [http-nio-80-exec-7] - Closing JDBC Connection [com.mysql.cj.jdbc.ConnectionImpl@6f2fd6f5]
    DEBUG [http-nio-80-exec-7] - Returned connection 1865406197 to pool.
```

服务端使用 sequ 得到订单编号元素，并将获取的订单信息与订单商品信息记录到对应的数据库表中。微信小程序效果如图 8-38 所示。

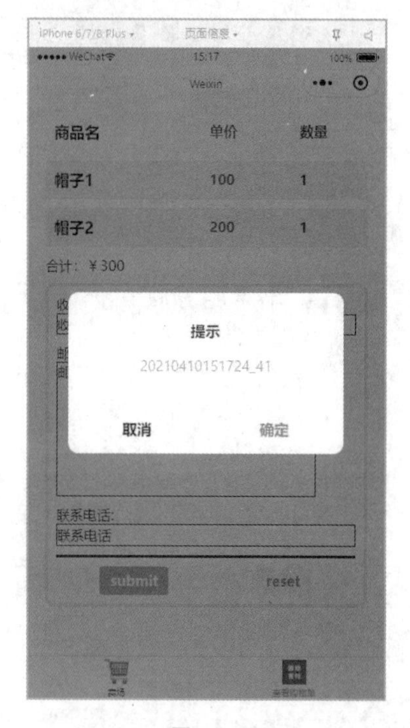

图 8-38

注意：

在实际微信小程序商城开发中，必须确保订单编号不重复，实现方式有多种。因为有的系统是微服务架构，有的系统是单体架构，有的系统是数据库分布式架构，所以实现方案各有差别，读者可根据各自项目的特点，酌情采用相应的技术方案。

订单提交后，提示框中给出的信息只有订单编号。在实际微信小程序商城开发中，还需要进行支付成功后的消息整合开发。

8.7 获取 openid

进行微信小程序开发时，一个重要的技术点是获取 openid。该知识点已在第 7 章中系统讲解过，这里不再赘述。

读者应注意的是，微信小程序中获取 openid 有多种技术解决方案。可以在微信小程序应用打开时请求服务器获取；可以在提交订单前获取；可以将需要获取 openid 的参数 code 连同订单一并提交到服务器。这取决于微信小程序设计的安全级别要求。

8.8 程序清单

程序清单分为 3 部分，即小程序端、服务端和数据库。

8.8.1 小程序端

小程序端的代码文件详情如下。

（1）app.js 程序代码如下。

```
App({

  /**
   * 当小程序初始化完成时，会触发 onLaunch（全局只触发一次）
   */
  onLaunch: function () {

  },

  /**
   * 当小程序启动，或从后台进入前台显示时，会触发 onShow
   */
  onShow: function (options) {

  },

  /**
   * 当小程序从前台进入后台时，会触发 onHide
   */
  onHide: function () {

  },

  /**
   * 当小程序发生脚本错误，或者 API 调用失败时，会触发 onError 事件并带上错误信息
   */
  onError: function (msg) {

  },
  url: 'http://localhost/shop/'
})
```

（2）app.json 代码如下。

```
{
  "pages": [
    "pages/shop/shop",
    "pages/shop1/shop1",
    "pages/queryOrder/queryOrder"
  ],
  "window": {
    "backgroundTextStyle": "light",
    "navigationBarBackgroundColor": "#fff",
    "navigationBarTitleText": "Weixin",
    "navigationBarTextStyle": "black"
  },
  "tabBar": {
    "list": [
      {
        "pagePath": "pages/shop/shop",
        "text": "卖场",
        "iconPath": "img/gwc.jpg",
        "selectedIconPath": "img/gwc.jpg"
      },
      {
        "pagePath": "pages/queryOrder/queryOrder",
        "text": "查看购物单",
        "iconPath": "img/pay.jpg",
        "selectedIconPath": "img/pay.jpg"
      }
    ]
  },
  "style": "v2",
  "sitemapLocation": "sitemap.json"
}
```

（3）小程序端无 app.wxss 代码。

（4）util.js 代码如下。

```
function formatTime(date) {
  var year = date.getFullYear()
  var month = date.getMonth() + 1
  var day = date.getDate()

  var hour = date.getHours()
  var minute = date.getMinutes()
  var second = date.getSeconds()


  return [year, month, day].map(formatNumber).join('/') + ' ' + [hour, minute, second].map(formatNumber).join(':')
}

function formatNumber(n) {
  n = n.toString()
  return n[1] ? n : '0' + n
}
```

```
module.exports = {
  formatTime: formatTime,
  requestData : requestData
}

function requestData(url,data,success){
  wx.request({
    url: url,
    data: data,
    header: {
      'Content-Type': 'application/json'
    },
    method: 'get',
    success: success
  })
}
```

（5）shop.js 代码如下。

```
var util = require('../../utils/util.js');
Page({

  /**
   * 页面的初始数据
   */
  data: {
    typeCode: 10,
  },
  selectTitle: function (event) {

    /**修改 title */
    this.setData({ typeCode: event.target.dataset.arg });

    this.showShangpin();

  },
  showShangpin: function () {

    util.requestData(
      this.data.url + 'Shop',
      {
        type: 'showShangPin',
        typeCode: this.data.typeCode
      },
      function (res) {
        //得到服务器端数据
        var pages = getCurrentPages();
        var page = pages[pages.length - 1];
        page.setData({ shangpin: res.data });
      }
    );
  },
  /**
   * 生命周期函数--监听页面加载
   */
  onLoad: function (options) {
```

```
    wx.setStorageSync('YigouShangpin', new Array());
    this.setData({ url: getApp().url });
    this.showShangpin();
  },

  /**
   * 生命周期函数--监听页面初次渲染完成
   */
  onReady: function () {

  },

  /**
   * 生命周期函数--监听页面显示
   */
  onShow: function () {

  },

  /**
   * 生命周期函数--监听页面隐藏
   */
  onHide: function () {

  },

  /**
   * 生命周期函数--监听页面卸载
   */
  onUnload: function () {

  },

  /**
   * 页面相关事件处理函数--监听用户下拉动作
   */
  onPullDownRefresh: function () {

  },

  /**
   * 页面上拉触底事件的处理函数
   */
  onReachBottom: function () {

  },

  /**
   * 用户单击右上角分享
   */
  onShareAppMessage: function () {

  },
  choose: function (event) {

    for (var x = 0; x < this.data.shangpin.length; x++) {
```

```
        if (event.currentTarget.dataset.shangpincode ==
this.data.shangpin[x].shangpincode) {
          wx.setStorageSync('chooseYigouShangpin', this.data.shangpin[x]);
        }
      }

      wx.navigateTo({ url: '/pages/shop1/shop1?shangpincode=' +
event.currentTarget.dataset.shangpincode });
    }
  })
```

（6）shop.json 代码如下。

```
  {
    "usingComponents": {}
  }
```

（7） shop.wxml 代码如下。

```
  <view class='title_view'>
    <button class='title' type='primary' size='mini' disabled='{{10==typeCode}}'
data-arg='10' bindtap="selectTitle">百货</button>
    <button class='title' type='primary' size='mini' disabled='{{20==typeCode}}'
data-arg='20' bindtap="selectTitle">餐厨</button>
    <button class='title' type='primary' size='mini' disabled='{{30==typeCode}}'
data-arg='30' bindtap="selectTitle">办公</button>
    <button class='title' type='primary' size='mini' disabled='{{40==typeCode}}'
data-arg='40' bindtap="selectTitle">游戏</button>
    <button class='title' type='primary' size='mini' disabled='{{50==typeCode}}'
data-arg='50' bindtap="selectTitle">动漫</button>
  </view>
  <scroll-view scroll-y='true' style='height:{{app.windowHeight-30}}px'>
    <view wx:for="{{shangpin}}" class='content'
data-shangpincode='{{item.shangpincode}}' bindtap='choose' wx:key="*this">
      <image class='shangpin_img' mode="aspectFit" src="{{url}}{{item.img}}" />
      <view class='context_text' style='width:{{app.windowWidth-150}}px'>
        <view style="height:50px">
          <text class='logo' wx:if='{{""!=item.logo}}'>{{item.logo}}</text>
          <text class="text">{{item.shangpinname}}</text>
        </view>
        <view style="height:20px;font-size:small">
          <text class='biaoqian' wx:if='{{""!=item.baoyou}}'>{{item.baoyou}}</text>
          <text class='biaoqian'
wx:if='{{""!=item.gouwuquan}}'>{{item.gouwuquan}}</text>
        </view>
        <view class="price">
          ¥: {{item.price}}
        </view>
      </view>
    </view>
  </scroll-view>
```

（8） shop.wxss 代码如下。

```
  .title_view{
   border-bottom-style:solid;
   border-bottom-color:blue;
```

```
  width: 100%;height:30px;
  background-color: #eee;
}

.title{
  margin-left: 1px;
  margin-right: 1px;
  height:29px;
}

.content{
  display: flex;
  margin: 3px 1px 3px 1px;
  border-bottom: 1px dashed black;
}

.shangpin_img{
  width: 100px;
  height: 100px;
}

.context_text{
  padding: 5px
}

.logo{
  background-color:red;
  color:yellow;
  font-size:small;
  margin-top: 1px;
  width: 2ex;
}

.text{
font-size:small
}

.biaoqian{
  color:red;
  border-color:red;
  border-style:solid;
  border-width: 1px;
  border-radius: 5px;
  margin: 1px;
  padding: 1px;
  font-size:small;
}

.price{
  color:red;
  font-size:small;
  height:20px;
  bottom:0;
}
```

（9）shop1.js 代码如下。

```
var util = require('../../utils/util.js');
Page({

  /**
   * 页面的初始数据
   */
  data: {
    shangpincode: null,

    ShangpinXiangqing: null,

    url: null,

    /**购买数 */
    count: 1,

    /**but_text */
    but_text: '立即购买',
  },

  /**
   * 生命周期函数--监听页面加载
   */
  onLoad: function (options) {
    this.setData({ shangpincode: options.shangpincode });
    this.setData({ url: getApp().url });

    util.requestData(this.data.url + 'Shop', { type: 'shopInfo', shangpincode:
options.shangpincode }, function (res) {
      var page = getCurrentPages()[getCurrentPages().length - 1];
      page.setData({ ShangpinXiangqing: res.data });
    });


  },

  /**
   * 生命周期函数--监听页面初次渲染完成
   */
  onReady: function () {

  },

  /**
   * 生命周期函数--监听页面显示
   */
  onShow: function () {

  },

  /**
   * 生命周期函数--监听页面隐藏
   */
  onHide: function () {
```

```
  },

  /**
   * 生命周期函数--监听页面卸载
   */
  onUnload: function () {

  },

  /**
   * 页面相关事件处理函数--监听用户下拉动作
   */
  onPullDownRefresh: function () {

  },

  /**
   * 页面上拉触底事件的处理函数
   */
  onReachBottom: function () {

  },

  /**
   * 用户单击右上角分享
   */
  onShareAppMessage: function () {

  },
  addition: function (event) {

    this.setData({ count: this.data.count * 1 + 1 });

  },
  subtraction: function (event) {

    if (0 > this.data.count * 1 - 1) {
      this.setData({ count: 0 });
    } else {
      this.setData({ count: this.data.count * 1 - 1 });
    }

  },
  buy: function () {

    var YigouShangpin = wx.getStorageSync('YigouShangpin');

    var flag = false;
    for (var x = 0; x < YigouShangpin.length; x++) {
      if (this.data.shangpincode == YigouShangpin[x].shangpincode) {
        YigouShangpin[x].count = this.data.count;
        flag = true;
        break;
      }
    }
```

```
    if (!flag) {
      var arg = {};
      arg = wx.getStorageSync('chooseYigouShangpin');
      arg.count = this.data.count;
      YigouShangpin.push(arg);
      this.setData({ count: arg.count });
    }

    if (0 == this.data.count) {
      this.setData({ but_text: '立即购买' });
    } else {
      this.setData({ but_text: '修改购买数量' });
    }

    YigouShangpin: wx.setStorageSync('YigouShangpin', YigouShangpin);

  }
})
```

（10）shop1.json 代码如下。

```
{
  "usingComponents": {}
}
```

（11）shop1.wxml 代码如下。

```
<view>
  <view wx:if='{{null!=ShangpinXiangqing}}'>
    <image class='shangpin_img'
src='{{url}}{{ShangpinXiangqing.shangpin.img}}'></image>
  </view>
  <view>
    <text class='logo'
wx:if='{{""!=ShangpinXiangqing.shangpin.logo}}'>{{ShangpinXiangqing.shangpin.logo}}</t
ext>
    <text class="shangpinname">{{ShangpinXiangqing.shangpin.shangpinname}}</text>
  </view>
  <view>
    <text class='text'>{{ShangpinXiangqing.shangpin.text }}</text>
  </view>
  <view>
    <text class='biaoqian'
wx:if='{{""!=ShangpinXiangqing.shangpin.baoyou}}'>{{ShangpinXiangqing.shangpin.baoyou}
}</text>
    <text class='biaoqian'
wx:if='{{""!=ShangpinXiangqing.shangpin.gouwuquan}}'>{{ShangpinXiangqing.shangpin.gouw
uquan}}</text>
  </view>
  <view style='font-size:small;'>
    <block wx:if='{{""!=ShangpinXiangqing.shangpin.zengpinbaozheng}}'>
      <icon size='15' type="success" />
      <text>{{ShangpinXiangqing.shangpin.zengpinbaozheng}}</text>
    </block>
    <block wx:if='{{""!=ShangpinXiangqing.shangpin.gouwuquan}}'>
      <icon size='15' type="success" />
```

```
        <text>{{ShangpinXiangqing.shangpin.gouwuquan}}</text>
      </block>
      <block wx:if='{{""!=ShangpinXiangqing.shangpin.quanguolianbao}}'>
        <icon size='15' type="success" />
        <text>{{ShangpinXiangqing.shangpin.quanguolianbao }}</text>
      </block>
      <block wx:if='{{""!=ShangpinXiangqing.shangpin.zengyunfeixian}}'>
        <icon size='15' type="success" />
        <text>{{ShangpinXiangqing.shangpin.zengyunfeixian }}</text>
      </block>
    </view>
    <view class="price">
      ￥: {{ShangpinXiangqing.shangpin.price}}
    </view>
    <view style='display: flex;'>
      <button size='mini' bindtap='subtraction'>-</button>
      <text>{{count}}</text>
      <button size='mini' bindtap='addition'>+</button>
      <button size='mini' type='primary' bindtap='buy'>{{but_text}}</button>
    </view>
    <view style='margin-top:10px'>
      <text style='font-weight:bold'>买家评价</text>
      <view wx:for='{{ShangpinXiangqing.shangpinlogs}}'
style='background-color:{{(0==index%2)?"antiquewhite":"burlywood"}};margin-top:10px;'
wx:for-item='log' wx:for-index='index' wx:key="*this">
        <view style='display: flex;'>
          <image mode='aspectFit' style='width:25px;height:15px'
src='{{url}}{{log.avatarUrl}}'></image>
          <text style='font-size:small;width:30%'>{{log.nickName}}</text>
          <view style='width:100%'></view>
          <text style='font-size:small;width:50%;'>{{log.createdate}}</text>
        </view>
        <view>
          <text>{{log.text}}</text>
        </view>
      </view>
    </view>
  </view>
```

（12）shop1.wxss 代码如下。

```
  .shangpin_img{
    width: 100%;
    margin: 0;
    padding: 0;
  }

  .logo{
    background-color:red;
    color:yellow;
    font-size:small;
    margin-top: 1px;
    width: 2ex;
  }

  .shangpinname{
    font-weight:bold
```

```
}

.text{

}

.biaoqian{
 color:red;
 border-color:red;
 border-style:solid;
 border-width: 1px;
 border-radius: 5px;
 margin: 1px;
 padding: 1px;
 font-size:small;
}

.price{
 color:red;
 font-size:large;
 height:20px;
 bottom:0;
}
```

（13）queryOrder.js 代码如下。

```
var util = require('../../utils/util.js');

Page({

  /**
   * 页面的初始数据
   */
  data: {

    YigouShangpin: null,
    all_price: 0,
    userInfo: null,
    url: ''
  },

  /**
   * 生命周期函数--监听页面加载
   */
  onLoad: function (options) {
    this.setData({ url: getApp().url });
  },

  /**
   * 生命周期函数--监听页面初次渲染完成
   */
  onReady: function () {
    wx.getUserInfo({
      success: function (res) {
        var pages = getCurrentPages();
        var page = pages[pages.length - 1];
        page.setData({ userInfo: res.userInfo });
```

```
      }
    })
  },

  /**
   * 生命周期函数--监听页面显示
   */
  onShow: function () {
    this.setData({ YigouShangpin: wx.getStorageSync('YigouShangpin') });
    var all_price = 0;
    for (var x = 0; x < this.data.YigouShangpin.length; x++) {
      all_price += this.data.YigouShangpin[x].price *
this.data.YigouShangpin[x].count * 100;
    }
    this.setData({ all_price: all_price / 100 });
  },

  /**
   * 生命周期函数--监听页面隐藏
   */
  onHide: function () {

  },

  /**
   * 生命周期函数--监听页面卸载
   */
  onUnload: function () {

  },

  /**
   * 页面相关事件处理函数--监听用户下拉动作
   */
  onPullDownRefresh: function () {

  },

  /**
   * 页面上拉触底事件的处理函数
   */
  onReachBottom: function () {

  },

  /**
   * 用户单击右上角分享
   */
  onShareAppMessage: function () {

  },
  formSubmit: function (event) {
    util.requestData(
      this.data.url + 'Shop',
      {
        type: 'saveOrder',
```

```
        userInfo: this.data.userInfo,
        userInfo_: event.detail.value,
        YigouShangpin: this.data.YigouShangpin,
      },
      function (res) {
        console.log(res.data)
        wx.showModal({
          title: '提示',
          content: '' + res.data.orderId,
          success: function (res) {
            console.log(res.data);
          }
        })
      });
  }
})
```

（14）queryOrder.json 代码如下。

```
{
  "usingComponents": {}
}
```

（15）queryOrder.wxml 代码如下。

```
<view>
  <view style="display: flex;">
    <view class="shangpinname">商品名</view>
    <view class="price">单价</view>
    <view class="count">数量</view>
  </view>

  <view style="display: flex;" wx:if="{{10==item.typecode}}"
class="typecode{{item.typecode}}" wx:for="{{YigouShangpin}}" wx:key="*this">
    <view class="shangpinname">{{item.shangpinname}}</view>
    <view class="price">{{item.price}}</view>
    <view class="count">{{item.count}}</view>
  </view>

  <view style="display: flex;" wx:if="{{20==item.typecode}}"
class="typecode{{item.typecode}}" wx:for="{{YigouShangpin}}" wx:key="*this">
    <view class="shangpinname">{{item.shangpinname}}</view>
    <view class="price">{{item.price}}</view>
    <view class="count">{{item.count}}</view>
  </view>

  <view style="display: flex;" wx:if="{{30==item.typecode}}"
class="typecode{{item.typecode}}" wx:for="{{YigouShangpin}}" wx:key="*this">
    <view class="shangpinname">{{item.shangpinname}}</view>
    <view class="price">{{item.price}}</view>
    <view class="count">{{item.count}}</view>
  </view>

  <view style="display: flex;" wx:if="{{40==item.typecode}}"
class="typecode{{item.typecode}}" wx:for="{{YigouShangpin}}" wx:key="*this">
    <view class="shangpinname">{{item.shangpinname}}</view>
    <view class="price">{{item.price}}</view>
```

```
      <view class="count">{{item.count}}</view>
    </view>

    <view style="display: flex;" wx:if="{{50==item.typecode}}"
class="typecode{{item.typecode}}" wx:for="{{YigouShangpin}}" wx:key="*this">
      <view class="shangpinname">{{item.shangpinname}}</view>
      <view class="price">{{item.price}}</view>
      <view class="count">{{item.count}}</view>
    </view>

    <view>合计: ¥{{all_price}}</view>

    <view class="form">

      <form bindsubmit="formSubmit">

        <view class="element">
          <label>收件人:
          <input type="text" name="sjr" placeholder="收件人" value="收件人"/></label>
        </view>

        <view class="element">
          <label>邮递地址:
          <textarea name="dz" placeholder="邮递地址" value="邮递地址"/></label>
        </view>

        <view class="element">
          <label>联系电话:
          <input name="dh" type="idcard" placeholder="联系电话" value="联系电话"/></label>
        </view>

        <view style="display: flex;   border-top-style:solid;
  border-top-color:blue;padding-top:10px;">
          <button type="primary" form-type="submit" size="mini" bindtap="submit"> submit
</button>
          <button type="warn"   form-type="reset"  size="mini" bindtap="reset" > reset
</button>
        </view>

      </form>

    </view>

  </view>
```

（16）queryOrder.wxss 代码如下。

```
  .shangpinname {
    font-weight: bold;
    font-size: large;
    width: 100%;
  }

  .price {
    color: red;
    font-weight: bold;
    padding-left: 1rem;
```

```
    padding-right: 1rem;
    width: 150px;
}

.count {
    padding-left: 10px;
    padding-right: 10px;
    color: blue;
    font-weight: bold;
    width: 150px;
}

.typecode10{background-color:antiquewhite;}
.typecode20{background-color:beige;}
.typecode30{background-color:gainsboro;}
.typecode40{background-color:wheat;}
.typecode50{background-color:darkseagreen;}

view{
  margin: 10px;
}

.element{
  margin-top: 10px;
}


.form{
  border-style:double;
  border-color:chartreuse;
  border-radius:10px;
}

input , textarea{
  border-style:solid;
  border-width: 1px;
}
```

8.8.2　服务端

服务端的程序主要有以下 5 个文件。

（1）Shop.java 代码如下。

```
package ajax;

import java.io.IOException;
import java.io.PrintWriter;
import java.text.SimpleDateFormat;
import java.util.Date;
import java.util.List;
import javax.servlet.ServletException;
import javax.servlet.annotation.WebServlet;
import javax.servlet.http.HttpServlet;
import javax.servlet.http.HttpServletRequest;
import javax.servlet.http.HttpServletResponse;
```

```
    import com.google.gson.Gson;
    import com.google.gson.GsonBuilder;
    import com.google.gson.reflect.TypeToken;
    import bean.ShangpinXiangqing;
    import generator.model.Shangpin;
    import generator.model.Shangpinlog;
    import generator.model.Userinfo;
    import generator.model.Xiaoshou;
    import server.ShopService;
    import util.WX_Util;

    @WebServlet("/Shop")
    public class Shop extends HttpServlet {

      protected void doGet(HttpServletRequest request, HttpServletResponse response)
throws ServletException, IOException {
        // TODO Auto-generated method stub
        doPost(request,response);
      }

      protected void doPost(HttpServletRequest request, HttpServletResponse response)
throws ServletException, IOException {

        request.setCharacterEncoding("UTF-8");
        response.setCharacterEncoding("UTF-8");

        PrintWriter out = response.getWriter();

        Gson gson = new GsonBuilder().setDateFormat("yyyy-MM-dd").create();

        String type= request.getParameter("type");
        System.out.println("type    "+type);

        if("showShangPin".equals(type)){

          Shangpin shangpin = new Shangpin();
          shangpin.setTypecode(request.getParameter("typeCode"));
          List<Shangpin> caipus = ShopService.queryShangpin(shangpin);
          out.print(gson.toJson(caipus));

        }else if("shopInfo".equals(type)){

          Shangpin shangpin = new Shangpin();
          shangpin.setShangpincode(request.getParameter("shangpincode"));
          List<Shangpin> caipus = ShopService.queryShangpin(shangpin);

          Shangpinlog shangpinlog = new Shangpinlog();
          shangpinlog.setShangpincode(request.getParameter("shangpincode"));
          List<Shangpinlog> shangpinlogs = ShopService.queryShangpinlog(shangpinlog);

          ShangpinXiangqing shangpinXiangqing = new
ShangpinXiangqing(caipus.get(0),shangpinlogs);

          out.print(gson.toJson(shangpinXiangqing));

        }else if("saveOrder".equals(type)){
```

```
        String orderId = WX_Util.getSequ("OrderId");

        Date date = new Date();
        SimpleDateFormat simpleDateFormat = new SimpleDateFormat("yyyyMMddHHmmss_");
        orderId = simpleDateFormat.format(date)+orderId;

        boolean flag = ShopService.saveOrder(
                    orderId,
                    date,
                    gson.fromJson(request.getParameter("userInfo"), Userinfo.class),
                    gson.fromJson(request.getParameter("userInfo_"),
Userinfo.class),
                    gson.fromJson(request.getParameter("YigouShangpin"), new
TypeToken<List<Xiaoshou>>(){}.getType()));
        if(flag){
          out.print("{\"flag\":\""+flag+"\",\"orderId\":\""+orderId+"\"}");
        }else{
          out.print("{\"flag\":\""+flag+"\"}");
        }

      }else if("".equals(type)){

      }else{
        out.println("type="+type);
      }

      out.flush();
      out.close();
    }

  }
```

（2）ShopService.java 代码如下。

```
  package server;

  import util.DBUtil;

  import java.util.Date;
  import java.util.List;
  import org.apache.ibatis.session.SqlSession;
  import generator.client.ShangpinMapper;
  import generator.client.ShangpinlogMapper;
  import generator.client.UserinfoMapper;
  import generator.client.XiaoshouMapper;
  import generator.model.Shangpin;
  import generator.model.ShangpinExample;
  import generator.model.ShangpinExample.Criteria;
  import generator.model.Shangpinlog;
  import generator.model.ShangpinlogExample;
  import generator.model.Userinfo;
  import generator.model.Xiaoshou;

  public class ShopService {

  public static List<Shangpin> queryShangpin(Shangpin shangpin){
```

```
        SqlSession sqlSession = DBUtil.getSqlSession();
        ShangpinMapper shangpinMapper = sqlSession.getMapper(ShangpinMapper.class);

        ShangpinExample example = new ShangpinExample();
        Criteria criteria = example.createCriteria();

    if(null!=shangpin.getShangpincode()){criteria.andShangpincodeEqualTo(shangpin.get
Shangpincode());}

    if(null!=shangpin.getTypecode()){criteria.andTypecodeEqualTo(shangpin.getTypecode
());}
        List<Shangpin> ShopList = shangpinMapper.selectByExample(example );

        sqlSession.close();

        return ShopList;

    }

    public static List<Shangpinlog> queryShangpinlog(Shangpinlog shangpinlog){

        SqlSession sqlSession = DBUtil.getSqlSession();
        ShangpinlogMapper shangpinlogMapper =
sqlSession.getMapper(ShangpinlogMapper.class);
        ShangpinlogExample example = new ShangpinlogExample();
        ShangpinlogExample.Criteria criteria = example.createCriteria();

    if(null!=shangpinlog.getShangpincode()){criteria.andShangpincodeEqualTo(shangpinl
og.getShangpincode());}

    if(null!=shangpinlog.getTypecode()){criteria.andTypecodeEqualTo(shangpinlog.getTy
pecode());}
        List<Shangpinlog> ShopList = shangpinlogMapper.selectByExample(example );

        sqlSession.close();

        return ShopList;

    }

    public static boolean saveOrder(String orderId, Date date, Userinfo userInfo,Userinfo
userInfo_, List<Xiaoshou> xiaoshous){

        boolean flag = false;

        SqlSession sqlSession = DBUtil.getSqlSession();

        try {
            UserinfoMapper userinfoMapper =
sqlSession.getMapper(UserinfoMapper.class);
            userInfo.setOrderid(orderId);
            userInfo.setCreatetime(date);
            userInfo.setDh(userInfo_.getDh());
            userInfo.setDz(userInfo_.getDz());
            userInfo.setSjr(userInfo_.getSjr());
```

```
            userinfoMapper.insertSelective(userInfo);

            XiaoshouMapper xiaoshouMapper =
sqlSession.getMapper(XiaoshouMapper.class);

            for(int x = 0 ; x < xiaoshous.size() ; x++){
                xiaoshous.get(x).setId(null);
                xiaoshous.get(x).setOrderid(orderId);
                xiaoshous.get(x).setCreatetime(date);
                xiaoshouMapper.insertSelective(xiaoshous.get(x));
            }

            sqlSession.commit();
            flag = true;
        } catch (Exception e) {
            sqlSession.rollback();
            e.printStackTrace();
        }

        sqlSession.close();

        return flag;
    }

    }
```

（3）DBUtil.java 代码如下。

```
    package util;

    import java.io.IOException;
    import java.io.InputStream;

    import org.apache.ibatis.io.Resources;
    import org.apache.ibatis.session.SqlSession;
    import org.apache.ibatis.session.SqlSessionFactory;
    import org.apache.ibatis.session.SqlSessionFactoryBuilder;

    public class DBUtil {

    private static SqlSessionFactory sqlSessionFactory = null;

    private DBUtil(){}

    public static SqlSession getSqlSession(){

        try {

            if(null == sqlSessionFactory){

                String resource = "cfg/mybatis-config.xml";

                InputStream inputStream = Resources.getResourceAsStream(resource);

                sqlSessionFactory = new SqlSessionFactoryBuilder().build(inputStream);

            }
```

```
            return sqlSessionFactory.openSession();
        } catch (IOException e) {
            e.printStackTrace();
            return null;
        }
    }
}
```

（4）WX_Util.java 代码如下。

```
package util;

import java.util.List;

import org.apache.ibatis.session.SqlSession;

import generator.client.SequMapper;
import generator.model.Sequ;
import generator.model.SequExample;
import generator.model.SequExample.Criteria;

public class WX_Util {

public synchronized static String getSequ(String col) {

    String str = null;

    SqlSession sqlSession = DBUtil.getSqlSession();

    SequMapper SequMapper = sqlSession.getMapper(SequMapper.class);

    SequExample example = new SequExample();
    Criteria criteria = example.createCriteria();
    criteria.andColNameEqualTo(col);
    List<Sequ> SequList = SequMapper.selectByExample(example);

    if (1 == SequList.size()) {

        Sequ Sequ = SequList.get(0);

        str = "" + (Integer.parseInt(Sequ.getCode()) + 1);

        Sequ.setCode(str);

        SequMapper.updateByPrimaryKey(Sequ);

        sqlSession.commit();

    }
```

```
        sqlSession.close();

        return str;

    }

}
```

（5）mybatis-config.xml 代码如下。

```
<?xml version="1.0" encoding="UTF-8" ?>
<!DOCTYPE configuration
  PUBLIC "-//mybatis.org//DTD Config 3.0//EN"
  "http://mybatis.org/dtd/mybatis-3-config.dtd">
<configuration>
<settings>
    <setting name="logImpl" value="LOG4J" />
</settings>
<environments default="development">
    <environment id="development">
        <transactionManager type="JDBC" />
        <dataSource type="POOLED">
            <property name="driver" value="com.mysql.cj.jdbc.Driver" />


            <property name="url"
value="jdbc:mysql://localhost:3306/shop?serverTimezone=UTC&amp;useSSL=true&amp;useUnic
ode=true&amp;characterEncoding=utf8" />
            <property name="username" value="root" />
            <property name="password" value="123456" />

            <!--<property name="useUnicode " value="true" />-->
            <!--<property name="characterEncoding " value="UTF-8" />-->
        </dataSource>
    </environment>
</environments>
<mappers>

    <mapper resource="generator/map/SequMapper.xml" />
    <mapper resource="generator/map/ShangpinMapper.xml" />
    <mapper resource="generator/map/ShangpinlogMapper.xml" />
    <mapper resource="generator/map/UserinfoMapper.xml" />
    <mapper resource="generator/map/XiaoshouMapper.xml" />


</mappers>
</configuration>
```

相关 JAR 文件包括以下几种。

- gson.jar。
- log4j-1.2.17.jar。
- log4j-api-2.0-rc1.jar。
- log4j-core-2.0-rc1.jar。
- mybatis.jar。

◆ mysql-connector-java-8.0.19.jar。

8.8.3 数据库

数据库相关代码如下。

```
/*
 Navicat MySQL Data Transfer

 Source Server         : localhost
 Source Server Type    : MySQL
 Source Server Version : 80019
 Source Host           : localhost:3306
 Source Schema         : shop

 Target Server Type    : MySQL
 Target Server Version : 80019
 File Encoding         : 65001

 Date: 10/04/2021 15:46:06
*/

SET NAMES utf8mb4;
SET FOREIGN_KEY_CHECKS = 0;

-- ----------------------------
-- Table structure for sequ
-- ----------------------------
DROP TABLE IF EXISTS `sequ`;
CREATE TABLE `sequ`  (
  `id` int(0) NOT NULL AUTO_INCREMENT,
  `col_name` varchar(50) CHARACTER SET utf8 COLLATE utf8_general_ci NOT NULL DEFAULT
'0' COMMENT '索引名',
  `code` varchar(50) CHARACTER SET utf8 COLLATE utf8_general_ci NOT NULL DEFAULT '0'
COMMENT '索引值',
  `text` varchar(50) CHARACTER SET utf8 COLLATE utf8_general_ci NULL DEFAULT NULL
COMMENT '说明',
  PRIMARY KEY (`id`) USING BTREE,
  UNIQUE INDEX `col_name`(`col_name`) USING BTREE
) ENGINE = InnoDB AUTO_INCREMENT = 1 CHARACTER SET = utf8 COLLATE = utf8_general_ci
ROW_FORMAT = Compact;

-- ----------------------------
-- Records of sequ
-- ----------------------------
INSERT INTO `sequ` VALUES (1, 'OrderId', '1', NULL);

-- ----------------------------
-- Table structure for shangpin
-- ----------------------------
DROP TABLE IF EXISTS `shangpin`;
CREATE TABLE `shangpin`  (
  `id` int(0) UNSIGNED NOT NULL AUTO_INCREMENT,
  `typecode` varchar(255) CHARACTER SET utf8 COLLATE utf8_general_ci NULL DEFAULT NULL
COMMENT '类型编号',
```

```
    `typename` varchar(255) CHARACTER SET utf8 COLLATE utf8_general_ci NULL DEFAULT NULL
COMMENT '类型名',
    `shangpincode` varchar(255) CHARACTER SET utf8 COLLATE utf8_general_ci NULL DEFAULT
NULL COMMENT '商品编号',
    `shangpinname` varchar(255) CHARACTER SET utf8 COLLATE utf8_general_ci NULL DEFAULT
NULL COMMENT '商品名',
    `logo` varchar(255) CHARACTER SET utf8 COLLATE utf8_general_ci NULL DEFAULT '' COMMENT
'标签',
    `price` decimal(10, 2) UNSIGNED NOT NULL COMMENT '价格',
    `img` varchar(255) CHARACTER SET utf8 COLLATE utf8_general_ci NULL DEFAULT NULL
COMMENT '图片',
    `baoyou` varchar(255) CHARACTER SET utf8 COLLATE utf8_general_ci NULL DEFAULT ''
COMMENT '包邮',
    `gouwuquan` varchar(255) CHARACTER SET utf8 COLLATE utf8_general_ci NULL DEFAULT ''
COMMENT '购物券',
    `zengpinbaozheng` varchar(255) CHARACTER SET utf8 COLLATE utf8_general_ci NULL
DEFAULT '' COMMENT '正品保证',
    `quanguolianbao` varchar(255) CHARACTER SET utf8 COLLATE utf8_general_ci NULL DEFAULT
'' COMMENT '全国联保',
    `zengyunfeixian` varchar(255) CHARACTER SET utf8 COLLATE utf8_general_ci NULL DEFAULT
'' COMMENT '赠运费险',
    `text` varchar(1000) CHARACTER SET utf8 COLLATE utf8_general_ci NOT NULL DEFAULT ''
COMMENT '简介',
    `odb` int(0) NOT NULL DEFAULT 0,
    `createdate` timestamp(0) NOT NULL DEFAULT CURRENT_TIMESTAMP(0) ON UPDATE
CURRENT_TIMESTAMP(0),
    PRIMARY KEY (`id`) USING BTREE
  ) ENGINE = InnoDB AUTO_INCREMENT = 1 CHARACTER SET = utf8 COLLATE = utf8_general_ci
COMMENT = '菜谱' ROW_FORMAT = Dynamic;

  -- ----------------------------
  -- Records of shangpin
  -- ----------------------------
  INSERT INTO `shangpin` VALUES (1, '10', 'tpyename10_1', '101', '帽子1', '热卖', 100.00,
'img/1001.jpg', '满1包邮', '满100.00送购物券', '正品', '全国联保', '赠运费险', '简介简介简介
简介简介简介简介简介简介简介简介', 0, '2021-04-05 18:28:08');
  INSERT INTO `shangpin` VALUES (2, '10', 'tpyename10_2', '102', '帽子2', NULL, 200.00,
'img/1002.jpg', '满2包邮', '', '正品', '全国联保', '赠运费险', '简介简介简介简介简介简介简介简
介简介简介简介', 0, '2021-04-05 18:28:09');
  INSERT INTO `shangpin` VALUES (3, '10', 'tpyename10_3', '103', '鞋3', '促销', 300.00,
'img/1003.jpg', '满3包邮', '满300.00送购物券', '正品', '全国联保', '赠运费险', '简介简介简介
简介简介简介简介简介简介简介简介', 0, '2021-04-05 18:28:15');
  INSERT INTO `shangpin` VALUES (4, '10', 'tpyename10_4', '104', '鞋4', NULL, 400.00,
'img/1004.jpg', '满4包邮', '满400.00送购物券', '正品', '全国联保', '赠运费险', '简介简介简介
简介简介简介简介简介简介简介简介', 0, '2021-04-05 18:28:17');
  INSERT INTO `shangpin` VALUES (6, '20', 'tpyename20_6', '206', '宝石6', '双十一', 600.00,
'img/2001.jpg', '满6包邮', '满600.00送购物券', '正品', '全国联保', '赠运费险', '简介简介简介
简介简介简介简介简介简介简介简介', 0, '2016-11-14 15:59:52');
  INSERT INTO `shangpin` VALUES (7, '20', 'tpyename20_7', '207', '宝石7', NULL, 700.00,
'img/2002.jpg', '满7包邮', '满700.00送购物券', '正品', '全国联保', '赠运费险', '简介简介简介
简介简介简介简介简介简介简介简介', 0, '2016-11-14 15:59:52');
  INSERT INTO `shangpin` VALUES (8, '20', 'tpyename20_8', '208', '宝石8', NULL, 800.00,
'img/2003.jpg', '满8包邮', '满800.00送购物券', '正品', '全国联保', '赠运费险', '简介简介简介
简介简介简介简介简介简介简介简介', 0, '2016-11-14 15:59:52');
  INSERT INTO `shangpin` VALUES (9, '20', 'tpyename20_9', '209', '宝石9', NULL, 900.00,
'img/2004.jpg', '满9包邮', '满900.00送购物券', '正品', '全国联保', '赠运费险', '简介简介简介
```

```
简介简介简介简介简介简介简介简介', 0, '2016-11-14 15:59:52');

    -- ----------------------------
    -- Table structure for shangpinlog
    -- ----------------------------
    DROP TABLE IF EXISTS `shangpinlog`;
    CREATE TABLE `shangpinlog`  (
      `id` int(0) UNSIGNED NOT NULL AUTO_INCREMENT,
      `typecode` varchar(255) CHARACTER SET utf8 COLLATE utf8_general_ci NULL DEFAULT NULL
COMMENT '类别编号',
      `typename` varchar(255) CHARACTER SET utf8 COLLATE utf8_general_ci NULL DEFAULT NULL
COMMENT '类别名',
      `shangpincode` varchar(255) CHARACTER SET utf8 COLLATE utf8_general_ci NULL DEFAULT
NULL COMMENT '商品编号',
      `shangpinname` varchar(255) CHARACTER SET utf8 COLLATE utf8_general_ci NULL DEFAULT
NULL COMMENT '商品名',
      `text` varchar(1000) CHARACTER SET utf8 COLLATE utf8_general_ci NOT NULL DEFAULT ''
COMMENT '菜品简介',
      `nick_name` varchar(255) CHARACTER SET utf8 COLLATE utf8_general_ci NULL DEFAULT NULL
COMMENT '昵称',
      `avatar_url` varchar(255) CHARACTER SET utf8 COLLATE utf8_general_ci NULL DEFAULT
NULL COMMENT '头像',
      `createdate` timestamp(0) NOT NULL DEFAULT CURRENT_TIMESTAMP(0) ON UPDATE
CURRENT_TIMESTAMP(0),
      PRIMARY KEY (`id`) USING BTREE
    ) ENGINE = InnoDB AUTO_INCREMENT = 1 CHARACTER SET = utf8 COLLATE = utf8_general_ci
ROW_FORMAT = Dynamic;

    -- ----------------------------
    -- Records of shangpinlog
    -- ----------------------------
    INSERT INTO `shangpinlog` VALUES (1, '10', 'tpyename10_1', '101', '哈雷彗星宝玉001',
'很好很好1', '昵称1', 'img/wx.jpg', '2016-11-11 14:23:54');
    INSERT INTO `shangpinlog` VALUES (2, '10', 'tpyename10_2', '102', '钻石001', '很好很
好2', '昵称2', 'img/wx.jpg', '2016-11-11 14:23:54');

    -- ----------------------------
    -- Table structure for userinfo
    -- ----------------------------
    DROP TABLE IF EXISTS `userinfo`;
    CREATE TABLE `userinfo`  (
      `id` int(0) UNSIGNED NOT NULL AUTO_INCREMENT,
      `orderid` varchar(255) CHARACTER SET utf8 COLLATE utf8_general_ci NOT NULL COMMENT
'订单id',
      `nick_name` varchar(255) CHARACTER SET utf8 COLLATE utf8_general_ci NULL DEFAULT NULL
COMMENT '昵称',
      `avatar_url` varchar(255) CHARACTER SET utf8 COLLATE utf8_general_ci NULL DEFAULT
NULL COMMENT '头像',
      `gender` varchar(255) CHARACTER SET utf8 COLLATE utf8_general_ci NULL DEFAULT NULL
COMMENT '性别',
      `province` varchar(255) CHARACTER SET utf8 COLLATE utf8_general_ci NULL DEFAULT NULL
COMMENT '省份\n\n',
      `city` varchar(255) CHARACTER SET utf8 COLLATE utf8_general_ci NULL DEFAULT NULL
COMMENT '城市',
      `sjr` varchar(255) CHARACTER SET utf8 COLLATE utf8_general_ci NULL DEFAULT NULL
COMMENT '收件人',
```

```
    `dz` varchar(255) CHARACTER SET utf8 COLLATE utf8_general_ci NULL DEFAULT NULL COMMENT '地址',
    `dh` varchar(255) CHARACTER SET utf8 COLLATE utf8_general_ci NULL DEFAULT NULL COMMENT '电话',
    `createtime` timestamp(0) NULL DEFAULT NULL ON UPDATE CURRENT_TIMESTAMP(0),
    PRIMARY KEY (`id`) USING BTREE
  ) ENGINE = InnoDB CHARACTER SET = utf8 COLLATE = utf8_general_ci ROW_FORMAT = Dynamic;

  -- ----------------------------
  -- Records of userinfo
  -- ----------------------------
  INSERT INTO `userinfo` VALUES (29, '20210410151724_41', '微信用户', 'https://thirdwx.qlogo.cn/mmopen/vi_32/POgEwh4mIHO4nibH0KlMECNjjGxQUq24ZEaGT4poC6icRiccVGKSyXwibcPq4BWmiaIGuGlicwxaQX6grC9VemZoJ8rg/132', '0', '', '', '收件人', '邮递地址', '联系电话', '2021-04-10 07:17:24');

  -- ----------------------------
  -- Table structure for xiaoshou
  -- ----------------------------
  DROP TABLE IF EXISTS `xiaoshou`;
  CREATE TABLE `xiaoshou`  (
    `id` int(0) UNSIGNED NOT NULL AUTO_INCREMENT,
    `orderid` varchar(255) CHARACTER SET utf8 COLLATE utf8_general_ci NULL DEFAULT NULL COMMENT '订单id',
    `count` int(0) NULL DEFAULT NULL COMMENT '数量',
    `typecode` varchar(255) CHARACTER SET utf8 COLLATE utf8_general_ci NULL DEFAULT NULL COMMENT '类别编号',
    `typename` varchar(255) CHARACTER SET utf8 COLLATE utf8_general_ci NULL DEFAULT NULL COMMENT '类别名',
    `shangpincode` varchar(255) CHARACTER SET utf8 COLLATE utf8_general_ci NULL DEFAULT NULL COMMENT '商品编号',
    `shangpinname` varchar(255) CHARACTER SET utf8 COLLATE utf8_general_ci NULL DEFAULT NULL COMMENT '商品名',
    `price` decimal(10, 2) NULL DEFAULT NULL COMMENT '价格',
    `img` varchar(255) CHARACTER SET utf8 COLLATE utf8_general_ci NULL DEFAULT NULL COMMENT '图片',
    `text` varchar(255) CHARACTER SET utf8 COLLATE utf8_general_ci NULL DEFAULT NULL COMMENT '说明',
    `createtime` timestamp(0) NULL DEFAULT NULL ON UPDATE CURRENT_TIMESTAMP(0),
    PRIMARY KEY (`id`) USING BTREE
  ) ENGINE = InnoDB AUTO_INCREMENT = 1 CHARACTER SET = utf8 COLLATE = utf8_general_ci ROW_FORMAT = Dynamic;

  -- ----------------------------
  -- Records of xiaoshou
  -- ----------------------------
  INSERT INTO `xiaoshou` VALUES (68, '20210410151724_41', 1, '10', 'tpyename10_1', '101', '帽子1', 100.00, 'img/1001.jpg', '简介简介简介简介简介简介简介简介简介简介', '2021-04-10 07:17:24');
  INSERT INTO `xiaoshou` VALUES (69, '20210410151724_41', 1, '10', 'tpyename10_2', '102', '帽子2', 200.00, 'img/1002.jpg', '简介简介简介简介简介简介简介简介简介简介', '2021-04-10 07:17:24');

  SET FOREIGN_KEY_CHECKS = 1;
```

第9章　小程序客服

微信小程序客服可实现客户端与客服人员之间的沟通。客户端的客服消息会话入口有以下两个。

- 微信小程序内。开发者在微信小程序内添加客服消息按钮组件，用户可在小程序内唤起客服会话页面，向微信小程序发送消息。
- 已使用过的小程序客服消息会聚合显示在微信会话"小程序客服消息"内，用户可以在小程序外查看历史客服消息，并向小程序客服发送消息。

微信小程序客服工具分为网页版与移动端两种，下面就来认识一下它们，并学会调用客服消息接口发送客服消息。

9.1　网页版小程序客服

使用小程序客服工具前，需要为小程序添加客服人员，并绑定微信账号。第一步，添加客服人员，如图9-1所示。第二步，绑定客服人员微信号，如图9-2所示。

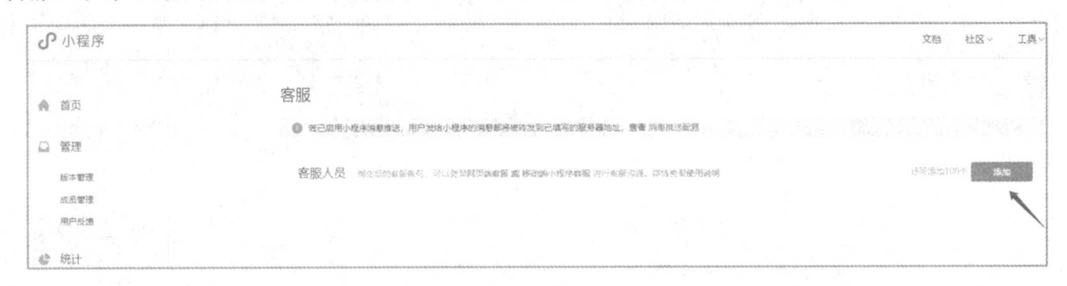

图9-1

图9-2

添加客户账号后，"客服人员"下方即可显示添加的客服人员信息，如图9-3所示。

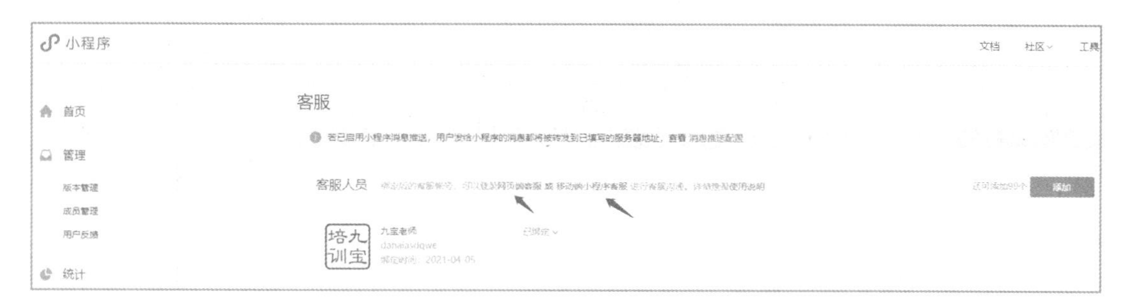

图 9-3

此时，用户可以单击“网页端客服”链接或“移动端小程序客服”链接，进入客服页面。

单击“网页端客服”链接后扫描二维码，即可登录网页版小程序客服工具，如图 9-4 所示。此时，对话框左上角会显示“0 人待回复”和“0 人待接入”。

图 9-4

下面为客户端创建“联系客服”按钮。

构建项目的“最小程序状态”，修改 WXML 文件，定义 button 组件，程序代码如下，开发页面如图 9-5 所示。

```
<button open-type="contact">联系客服</button>
```

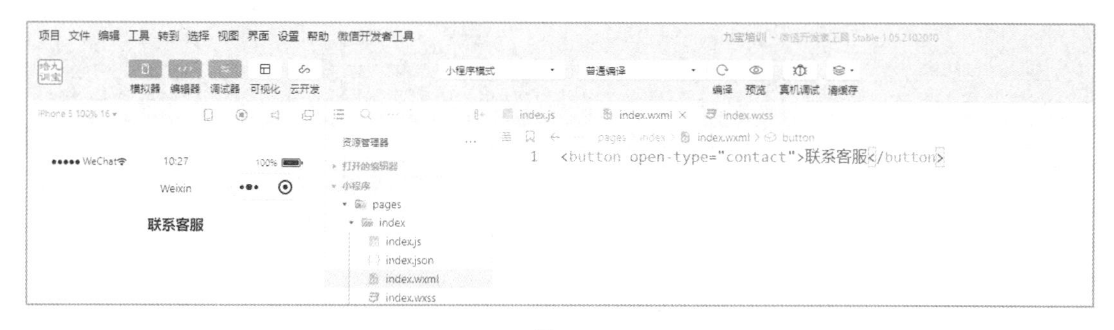

图 9-5

在菜单功能区先单击“真机调试”按钮，然后在“自动真机调试”选项卡下选中“启动手机端自动真机调试”单选按钮，如图 9-6 所示。

图 9-6

此时，微信小程序手机端的页面效果如图 9-7 所示。当用户单击“联系客服”按钮时，会跳转到客服聊天页面，如图 9-8 所示。此时用户可发送 text 类型的消息给客服人员，如图 9-9 所示。

图 9-7

图 9-8

图 9-9

收到用户发来的消息后，网页客服工具会提示“待接入 1”，客服人员选择要接入的用户，即可开始对话聊天，如图 9-10、图 9-11 所示。

图 9-10

图 9-11

9.2　移动端小程序客服

移动端小程序客服工具与网页版客服工具的操作类似。

首先，在如图 9-12 所示页面中单击“移动端小程序客服”链接，扫描二维码，登录移动端小程序客服工具。第一步，进入小程序，如图 9-13 所示。第二步，设置允许向你发送小程序服务通知，如图 9-14 所示。第三步，显示客服聊天页面，如图 9-15 所示。

此时客户端发来消息，客服工具收到消息即可进行对话聊天，如图 9-16 所示。

图 9-12

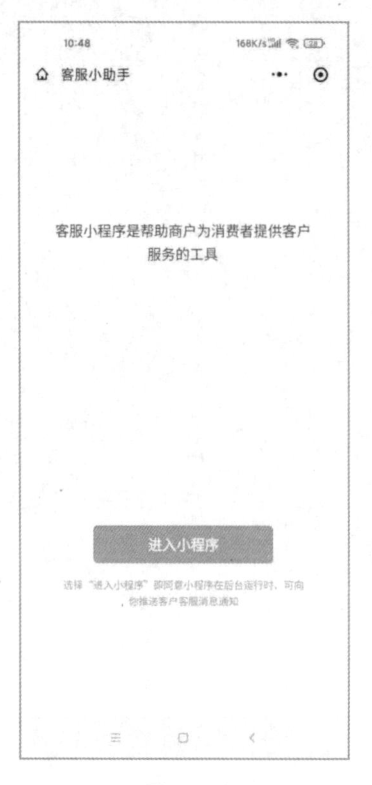

图 9-13

10:48

客服小助手

客服小程序是帮助商户为消费者提供客户服务的工具

客服小助手 申请

向你发送小程序服务通知

拒绝 允许

图 9-14

图 9-15

图 9-16

9.3 调用客服消息接口发送消息

对于调用客服消息接口发送消息功能，需要通过编程来实现。

新建服务端应用，WX_Interface 类用于接收微信小程序服务器发送的请求信息，代码如下。

```
    package util;

    import java.io.IOException;
    import javax.servlet.ServletException;
    import javax.servlet.annotation.WebServlet;
    import javax.servlet.http.HttpServlet;
    import javax.servlet.http.HttpServletRequest;
    import javax.servlet.http.HttpServletResponse;

    import wx.bean.request.WX_In_text;

    @WebServlet("/WX_Interface")
    public class WX_Interface extends HttpServlet {
    protected void doGet(HttpServletRequest request, HttpServletResponse response) throws
ServletException, IOException {

        response.getWriter().print(WX_Util.validate(request));
    }

    protected void doPost(HttpServletRequest request, HttpServletResponse response)
throws ServletException, IOException {

        request.setCharacterEncoding("UTF-8");
        response.setCharacterEncoding("UTF-8");

        String requestStr = WX_Util.getStringInputStream(request);

        System.out.println(requestStr);

        WX_In_text in_text = new WX_In_text(requestStr);

        System.out.println(in_text.getContent());

        response.getWriter().print("success");
    }

    }
```

WX_ARG 类用于提供参数，代码如下。

```
    package util;

    public class WX_ARG {
        public static String appID = "";
        public static String appsecret = "";
    }
```

AppID 和 AppSecret 是微信小程序 ID 和密钥，获取方式如图 9-17 所示。

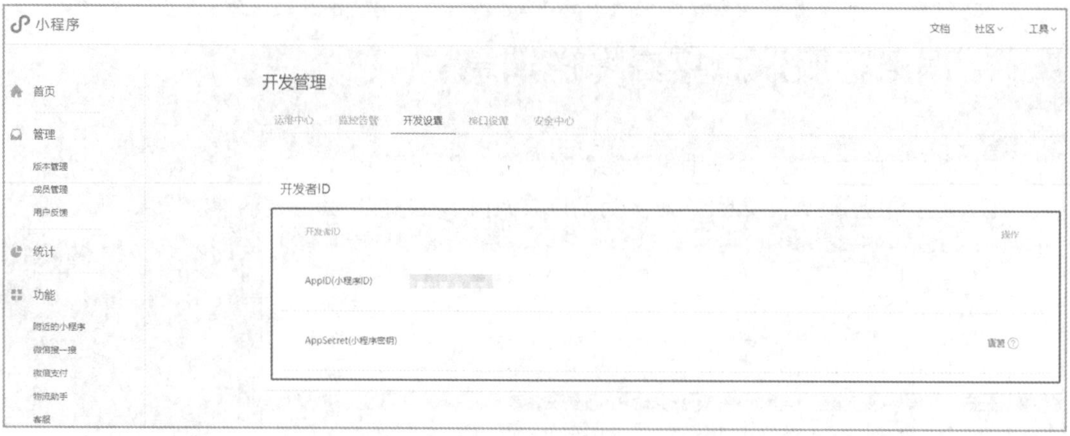

图 9-17

WX_Util 类用于提供计算的工具方法，代码如下。

```
package util;

import java.io.BufferedReader;
import java.io.IOException;
import java.io.InputStreamReader;
import java.io.OutputStreamWriter;
import java.io.StringReader;
import java.net.HttpURLConnection;
import java.net.URL;
import java.security.MessageDigest;
import java.util.Arrays;
import javax.servlet.http.HttpServletRequest;
import javax.xml.parsers.DocumentBuilder;
import javax.xml.parsers.DocumentBuilderFactory;

import org.w3c.dom.Document;
import org.w3c.dom.Element;
import org.w3c.dom.NodeList;
import org.xml.sax.InputSource;

import com.google.gson.Gson;
import wx.bean.AccessToken;
import wx.bean.response.WX_Out;

public class WX_Util {
public static String validate(HttpServletRequest request) {
     String signature = request.getParameter("signature");
     String timestamp = request.getParameter("timestamp");
     String nonce = request.getParameter("nonce");
     String echostr = request.getParameter("echostr");
     String token = "jiubao";
     String str = "";
     try {
          if (null != signature) {
               String[] ArrTmp = { token, timestamp, nonce };
               Arrays.sort(ArrTmp);
               StringBuffer sb = new StringBuffer();
               for (int i = 0; i < ArrTmp.length; i++) {
                    sb.append(ArrTmp[i]);
               }
               MessageDigest md = MessageDigest.getInstance("SHA-1");
               byte[] bytes = md.digest(new String(sb).getBytes());
               StringBuffer buf = new StringBuffer();
               for (int i = 0; i < bytes.length; i++) {
```

```
                if (((int) bytes[i] & 0xff) < 0x10) {
                    buf.append("0");
                }
                buf.append(Long.toString((int) bytes[i] & 0xff, 16));
            }
            if (signature.equals(buf.toString())) {
                str = echostr;
            }
        }
    } catch (Exception e) {
        e.printStackTrace();
    }
    return str;
}

public static String getStringInputStream(HttpServletRequest request) {
    InputStreamReader reader = null;
    BufferedReader breader = null;
    StringBuffer strb = new StringBuffer();
    try {
        reader = new InputStreamReader(request.getInputStream(), "UTF-8");
        breader = new BufferedReader(reader);
        String str = null;
        while (null != (str = breader.readLine())) {
            strb.append(str);
        }
    } catch (IOException e) {
        e.printStackTrace();
    }
    if (null != reader) {
        try {
            reader.close();
        } catch (IOException e) {
            e.printStackTrace();
        }
    }
    if (null != breader) {
        try {
            breader.close();
        } catch (IOException e) {
            e.printStackTrace();
        }
    }
    return strb.toString();
}

public static String getXMLCDATA(String requestStr, String tagName) {
    try {
        DocumentBuilderFactory dbf = DocumentBuilderFactory.newInstance();
        DocumentBuilder db = dbf.newDocumentBuilder();
        StringReader sr = new StringReader(requestStr);
        InputSource is = new InputSource(sr);
        Document document = db.parse(is);
        Element root = document.getDocumentElement();
        NodeList nodeList = root.getElementsByTagName(tagName);
        if(0!=nodeList.getLength()){
            return root.getElementsByTagName(tagName).item(0).getTextContent();

        }else{
            return "";
        }
    } catch (Exception e) {
        e.printStackTrace();
        return "";
    }
```

```
    }
}
```

在“客服”页面单击“消息推送配置”链接，如图 9-18 所示，启用消息推送功能并设置消息推送配置，如图 9-19、图 9-20 所示。设置完毕后将显示配置信息，如图 9-21 所示。

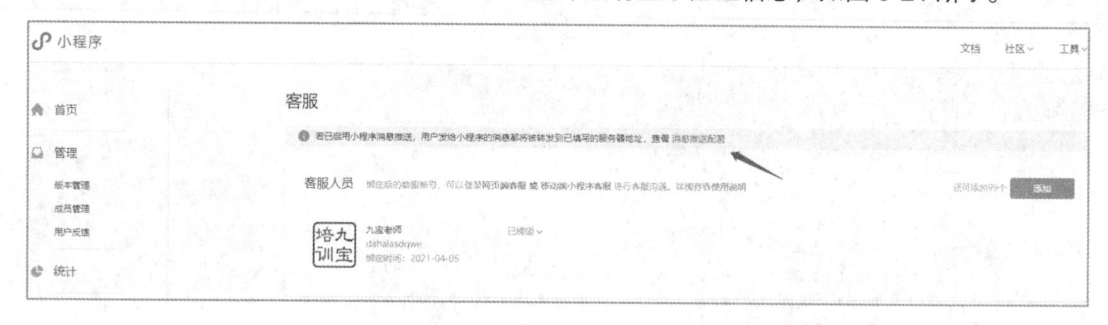

图 9-18

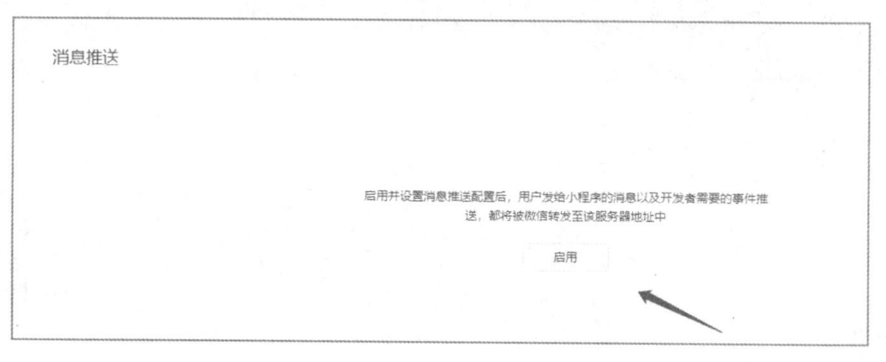

图 9-19

消息推送配置

填写的URL需要正确响应微信发送的Token验证，填写说明请阅读消息推送服务器配置指南

URL(服务器地址)　https:// /wx/WX_Interface

必须以http://或https://开头，分别支持80端口和443端口

Token(令牌)　jiubao

必须为英文或数字，长度为3-32字符

EncodingAESKey (消息加密密钥)　6UHkxeYRQJG6xm1U8zmVZD07pXlrO8i3mTmWsHNooYE 43/43　随机生成

消息加密密钥由43位字符组成，字符范围为A-Z,a-z,0-9

消息加密方式　请根据业务需要，选择消息加密方式，启用后将立即生效

明文模式（不使用消息体加解密功能，安全系数较低）

兼容模式（明文、密文将共存，方便开发者调试和维护）

安全模式（推荐）（消息包为纯密文，需要开发者加密和解密，安全系数高）

数据格式　请选择微信与开发者服务器间传输的数据格式

JSON　XML

提交

图 9-20

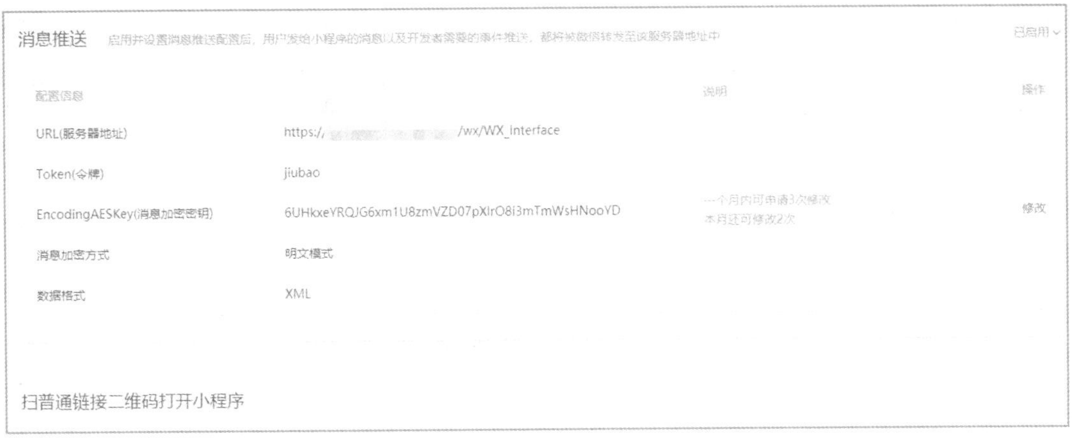

消息推送 启用并设置消息推送配置后，用户发给小程序的消息以及开发者需要的事件推送，都将被微信转发至该服务器地址中 已启用

配置信息		说明	操作
URL(服务器地址)	https:/ /wx/WX_interface		
Token(令牌)	jiubao		
EncodingAESKey(消息加密密钥)	6UHkxeYRQJG6xm1U8zmVZD07pXlrO8i3mTmWsHNooYD	一个月内可申请3次修改 本月还可修改2次	修改
消息加密方式	明文模式		
数据格式	XML		

扫普通链接二维码打开小程序

图 9-21

9.4 消息转发给客服人员

用户向公众号发送消息时，微信服务器会将消息以 POST 方式发送到开发者填写的 url 上。如果希望将消息同步转发到客服系统，需要在响应包中返回 MsgType 为 transfer_customer_service 的消息，微信服务器收到响应后会把当次发送的消息转发至客服系统。

对应的程序代码如下，其中的参数说明如表 9-1 所示。

```
<xml>
  <ToUserName><![CDATA[touser]]></ToUserName>
  <FromUserName><![CDATA[fromuser]]></FromUserName>
  <CreateTime>1399197672</CreateTime>
  <MsgType><![CDATA[transfer_customer_service]]></MsgType>
</xml>
```

表 9-1

参　数	必　须	描　述
ToUserName	是	接收方账号（收到的 OpenID）
FromUserName	是	开发者微信号
CreateTime	是	消息创建时间 （整型）
MsgType	是	transfer_customer_service

9.5 消息转发给指定客服人员

在多客服系统中，通常会有多个客服人员登录客服后台，并同时开启自动接入功能，等待用户接入。当有客户消息进入时，多客服系统会将客户随机分配给其中一个客服人员。如果希望将某个客户的消息转给指定的客服来接待，可在返回 transfer_customer_service 消息时附上 TransInfo 信息，为其指定一个客服账号。

▶ **注意：**

如果指定的客服没有接入能力，如不在线、没有开启自动接入功能或者自动接入人数已满，则该用户仍然会被直接接入指定客服，不再通知其他客服，也不会被其他客服接待。因此，建议在指定客服时，先查询客服的接入能力（即获取在线客服接待信息接口），指定有能力接入的客服，以保证客户能够及时得到服务。

对应的程序代码如下，其中的参数说明如表 9-2 所示。

```
<xml>
  <ToUserName><![CDATA[touser]]></ToUserName>
  <FromUserName><![CDATA[fromuser]]></FromUserName>
  <CreateTime>1399197672</CreateTime>
  <MsgType><![CDATA[transfer_customer_service]]></MsgType>
  <TransInfo>
    <KfAccount><![CDATA[test1@test]]></KfAccount>
  </TransInfo>
</xml>
```

表 9-2

参　数	必　须	描　述
ToUserName	是	接收方账号（收到的 OpenID）
FromUserName	是	开发者微信号
CreateTime	是	消息创建时间（整型）
MsgType	是	transfer_customer_service
KfAccount	是	指定会话接入的客服账号

获取客服的基本信息，可以使用 GET 请求方式：https://api.weixin.qq.com/cgi-bin/customservice/getkflist?access_token=ACCESS_TOKEN，请求结果如下。

```
{
  "kf_list": [
    {
      "kf_account": "kf2000@gh_5ff181ffcac8",
      "kf_headimgurl": "http://wx.qlogo.cn/mmhead/PiajxSqBRaEJ0RTMzlw7ibrfuzicOtPm
kv5sIbmdkAOxyIIgKEds0cB2A/0",
      "kf_id": 2003,
      "kf_nick": "九宝老师",
      "kf_wx": "dahaiasdqwe"
    }
  ]
}
```

其中的参数说明如表 9-3 所示。

表 9-3

参　数	说　明
kf_account	完整的客服账号，格式：账号前缀@公众号/微信号
kf_nick	客服昵称
kf_id	客服编号
kf_headimgurl	客服头像
kf_wx	如果客服账号已绑定客服人员微信号，此处显示微信号

access_token 是全局唯一接口调用凭据，可以使用 GET 请求方式：https://api.weixin.

qq.com/cgi-bin/token?grant_type=client_credential&appid=APPID&secret=APPSECRET。

参数说明如表 9-4 所示。

表 9-4

参　数	必　须	说　明
grant_type	是	获取 access_token 填写 client_credential
appid	是	第三方用户唯一凭证
secret	是	第三方用户唯一凭证密钥，即 appsecret

9.6　发送客服消息

微信小程序允许发送的消息类型有很多种，下面以发送 text 文本消息为例进行讲解。

修改 WX_Util 类，定义 public static AccessToken getAccessToken()与 public static void sendServer(AccessToken accessToken, WX_Out wx_Out)，程序代码如下。

```
    public static AccessToken getAccessToken() {
         AccessToken accessToken = null;
         InputStreamReader reader = null;
         BufferedReader breader = null;
         try {
              URL url = new URL("https://api.weixin.qq.com/cgi-bin/token?grant_type
=client_credential&appid=" + WX_ARG.appID + "&secret=" + WX_ARG.appsecret);
              HttpURLConnection connection = (HttpURLConnection) url.openConnection();
              connection.connect();
              reader = new InputStreamReader(connection.getInputStream());
              breader = new BufferedReader(reader);
              String str = null;
              StringBuffer strb = new StringBuffer();
              while (null != (str = breader.readLine())) {
                   strb.append(str);
              }
              Gson gson = new Gson();
              accessToken = gson.fromJson(strb.toString(), AccessToken.class);
         } catch (Exception e) {
              e.printStackTrace();
         }
         try {
              breader.close();
              reader.close();
         } catch (IOException e) {
              e.printStackTrace();
         }
         return accessToken;
    }

    public static void sendServer(AccessToken accessToken, WX_Out wx_Out) {
         try {
              URL url = new URL(https://api.weixin.qq.com/cgi-bin/message/custom/send?
access_token= + accessToken.getAccess_token());
              HttpURLConnection connection = (HttpURLConnection) url.openConnection();
              connection.setDoOutput(true);
              connection.connect();
              OutputStreamWriter writer = new
OutputStreamWriter(connection.getOutputStream());
              writer.write(wx_Out.getServerStr());
```

```
            writer.flush();
            InputStreamReader reader = new
InputStreamReader(connection.getInputStream());
            BufferedReader breader = new BufferedReader(reader);
            StringBuffer strb = new StringBuffer();
            String str = null;
            while (null != (str = breader.readLine())) {
                strb.append(str);
            }
            System.out.println(strb.toString());
            writer.close();
            reader.close();
            breader.close();
            connection.disconnect();
        } catch (Exception e) {
            e.printStackTrace();
        }
    }
```

代码解析如下。

- public static AccessToken getAccessToken()方法用于获取 access_token。注意，access_token 必须要缓存。
- public static void sendServer(AccessToken accessToken, WX_Out wx_Out)方法用于发送客服消息。

WX_Out 类用于封装客户消息，代码如下。

```
package wx.bean.response;
public interface WX_Out {
     public String getServerStr();
}
```

WX_Out_text 类用于封装 text 类型的客户消息，代码如下。

```
package wx.bean.response;
/* @author 九宝*/
 * @author Administrator
 */
public class WX_Out_text implements WX_Out {

private String touser = null;
private String content = null;

public String getServerStr() {
     StringBuffer strb = new StringBuffer();
     strb.append(" { ");
     strb.append("    \"touser\":\""+this.getTouser()+"\", ");
     strb.append("    \"msgtype\":\"text\", ");
     strb.append("    \"text\": ");
     strb.append("    { ");
     strb.append("        \"content\":\""+this.getContent()+"\" ");
     strb.append("    } ");
     strb.append(" } ");
     return strb.toString();
}

public String getTouser() {
     return touser;
}

public void setTouser(String touser) {
     this.touser = touser;
```

```
}

public String getContent() {
    return content;
}

public void setContent(String content) {
    this.content = content;
}

}
```

Test 类用于发送客服消息，代码如下。

```
import util.WX_Util;
import wx.bean.response.WX_Out_text;

public class Test {

  public static void main(String[] args) {
     // TODO Auto-generated method stub

     WX_Out_text out_text = new WX_Out_text();
     out_text.setContent("九宝 2326321088");
     out_text.setTouser("oZ8L60KpGiHLNHjbo2Feio5ZvNtg");

     WX_Util.sendServer(WX_Util.getAccessToken(), out_text);
  }

}
```

执行测试，调试器 Console 打印以下信息。

```
{"errcode":0,"errmsg":"ok"}
```

微信手机端的显示效果如图 9-22 所示。

图 9-22